ABOUT ISLAND PRESS

Island Press is the only nonprofit organization in the United States whose principal purpose is the publication of books on environmental issues and natural resource management. We provide solutions-oriented information to professionals, public officials, business and community leaders, and concerned citizens who are shaping responses to environmental problems.

Since 1984, Island Press has been the leading provider of timely and practical books that take a multidisciplinary approach to critical environmental concerns. Our growing list of titles reflects our commitment to bringing the best of an expanding body of literature to the environmental community throughout North America and the world.

Support for Island Press is provided by the Agua Fund, The Geraldine R. Dodge Foundation, Doris Duke Charitable Foundation, The Ford Foundation, The William and Flora Hewlett Foundation, The Joyce Foundation, Kendeda Sustainability Fund of the Tides Foundation, The Forrest & Frances Lattner Foundation, The Henry Luce Foundation, The John D. and Catherine T. MacArthur Foundation, The Marisla Foundation, The Andrew W. Mellon Foundation, Gordon and Betty Moore Foundation, The Curtis and Edith Munson Foundation, Oak Foundation, The Overbrook Foundation, The David and Lucile Packard Foundation, Wallace Global Fund, The Winslow Foundation, and other generous donors.

The opinions expressed in this book are those of the authors and do not necessarily reflect the views of these foundations.

MOUNTAIN GOATS

MOUNTAIN GOATS

Ecology, Behavior, and Conservation
of an Alpine Ungulate

Marco Festa-Bianchet
Steeve D. Côté

ISLANDPRESS
Washington · Covelo · London

Library of Congress Cataloging-in-Publication Data

Festa-Bianchet, Marco.
  Mountain goats : ecology, behavior, and conservation of an alpine ungulate / Marco Festa-Bianchet, Steeve Côté.
    p. cm.
  Includes bibliographical references and index.
  ISBN 978-1-59726-170-8 (hardcover : alk. paper) --
  ISBN 978-1-59726-171-5 (pbk. : alk. paper)
  1. Mountain goat. I. Côté, Steeve D. II. Title.
  QL737.U53F48 2007
  639.97'96475--dc22
                                                              2007025958

Printed on recycled, acid-free paper ✪

Manufactured in the United States of America

10 9 8 7 6 5 4 3 2 1

À Wendy et Mélanie,
qui ont toujours accepté et respecté nos longs séjours
sur la montagne . . .

E per Alberto,
a cui sarebbe piaciuto vedere le capre.

Contents

Acknowledgments

Over sixteen years, many people helped us continue our research on mountain goats, and several funding agencies and organizations supported us financially or helped us logistically. First and foremost, we wish to thank the many collaborators, students, and assistants who helped us collect data in the field. These people endured snowstorms in July, mechanical breakdowns, questionable food, pesky marmots, and the presence of grizzly bears to observe, capture, or otherwise study mountain goats. We hope all have fond memories of Caw Ridge. In alphabetical order, they are: Chantal Beaudoin, Félix Boulanger, Étienne Cardinal, Guillaume Côté, Étienne Drouin, Donald Dubé, Catherine Gagnon, Yanick Gendreau, Sandra Hamel, Dave Hildebrand, Mélina Houle, Paul Jones, Sandro Lovari, Fanie Pelletier, Alberto Peracino, Sabrina Popp, Sébastien Rioux, Giorgia Romeo, John Russell, Ken Seidel, Geneviève Simard, Martin Urquhart, Lucie Vallières, Vanessa Viera, and Sébastien Wendenbaum. We wish to single out Martin Urquhart, who ensured that our first critical years of fieldwork were successful, Yanick Gendreau, who contributed much effort and enthusiasm to collecting data from 1998 to 2002, and Sandra Hamel, whose passion for mountain goats and for Caw Ridge defies description, and who was the main force behind this research from 2002 through 2006.

The real work in ecological research is done by graduate students, and the Caw Ridge study was no exception. We are thankful to have had so many enthusiastic and resourceful graduate students, the secret weapon of this research program. Many have gone on to successful careers in wildlife conservation or research, and we are proud of their achievements. In chronological order, they are François Fournier, Martine Haviernick, Alejandro Gonzalez-Voyer, Yanick Gendreau, Sandra Hamel, and Julien Mainguy.

The logistics of the Caw Ridge study are complex, as is to be expected of any research program in remote areas with difficult access and unreliable communication lines. We thank all those that made our life easier in the field by providing logistical support, helped us in the laboratory, or generally got us out of trouble: Bill Allen, Janet Ficht, Steven Cross, Mike Ewald, Dave Hobson, Dave McKenna, Bertrand Mercier, Kirby Smith, and Shane Ramstead. Institutional logistical support was provided by the Alberta Forest Service, Grande Cache Correctional Center (we are very grateful for the construction of our traps and the field cabin), Renewable Resources Consultants, and Smoky River Coal, Ltd.

As we wrote various drafts of this book, we received many wise and constructive comments from several colleagues, including Tim Coulson, Jean-Michel Gaillard, Sandra Hamel, Wendy King, Julien Mainguy, Fanie Pelletier, Cliff Rice, Kathreen Ruckstuhl, and Kirby Smith. We thank Dave Coltman, Tim Coulson, Jean-Michel Gaillard, Jon Jorgenson, and Kirby Smith for ideas and discussions that helped shape our thinking about mountain ungulates and their conservation.

We acknowledge the pivotal role played by Kirby Smith in the Caw Ridge Mountain goat study. Kirby first suggested that we choose Caw Ridge as a study area, set up the initial capture operation, helped us define the goals of the study, and provided logistic support on innumerable occasions. His unswerving dedication to wildlife conservation and to the study of mountain goats was among the main assets of this research program. His good humor, hospitality, and knowledge were always much appreciated.

We were able to carry on this long-term study because we were financially supported by agencies that recognize the importance of fundamental research on wild animals. Our study was generously supported by the Natural Sciences and Engineering Research Council of Canada (NSERC, which provided operating and equipment grants to us and scholarships to some of our students), the Fonds Québécois de la Recherche sur la Nature et les Technologies, Alberta Fish and Wildlife, Alberta Conservation Association (ACA), Alberta Recreation, Parks and Wildlife Foundation, Alberta Wildlife Enhancement Fund, Challenge Grants in Biodiversity (ACA), International Order of Rocky Mountain Goats, Rocky Mountain Goat Foundation, the Université de Sherbrooke, and Université Laval. We thank Alberta Fish and Wildlife and the British Columbia Ministry of the Environment for supporting the publication of this book.

CHAPTER 1

Ecological Questions, Conservation Challenges, and Long-Term Research

The conservation of biodiversity and the management of wildlife require an understanding of the basic ecology of animals. That deceptively simple statement conceals a difficult problem, because understanding the "basic ecology" of species demands years of research. The processes that affect population dynamics of large mammals often develop over many years and cannot be understood without long-term monitoring. Important events (such as forest fires, extreme weather, or the appearance of new predators, competitors, or diseases) may have drastic long-term effects on population dynamics but they may be too rare to be detected, let alone quantified, by a few years of research. In addition, the many factors affecting a species' abundance seldom act in isolation. Instead, interactions between factors are commonplace: for example, body mass may affect survival only at high population density (Festa-Bianchet et al. 1997), and the impacts of a harsh winter may vary substantially according to a population's age structure (Coulson et al. 2000). Similar complex relationships affect the consequences of different harvest levels, which can be radically different according to the sex–age composition of the population and of harvested animals.

Consequently, an in-depth understanding of ungulate ecology requires data collected over many years and can best be served by long-term studies that seek to answer fundamental questions: What affects population size? What factors determine age- and sex-specific mortality? How do individuals differ in their ability to contribute to population recruitment, and why do those differences exist?

A long-term approach to the study of the ecology and conservation of large herbivores is particularly appropriate because of their longevity and complex population structure. An individual can experience varying levels of environmental conditions over its lifetime. Consequently, the reproductive strategy of large herbivores likely evolved in response to the range of environmental conditions that an individual may encounter over its lifetime. The complex population structure of large herbivores, sometimes involving a dozen or more coexisting cohorts, means that the population present today is the result of processes and events that took place over the previous decade and will affect population dynamics over the next one. It is therefore essential that management programs to conserve large herbivores, including those that involve some harvest, be mindful of the differences among individuals. In addition, the consequences of conservation actions (or of harvest strategies) can persist over many years. Because of the importance of differences among individuals and of processes occurring over several years, biological knowledge useful for conservation of large herbivores can therefore be best obtained by long-term monitoring of known individuals within a population.

Public finances typically sustain fundamental ecological studies. Our mountain goat study is no exception. In addition to producing novel information, ecological studies have an obligation to clearly communicate the applied implications of their results. The conservation of biodiversity requires long-term research, and long-term research should make a contribution to conservation. We will use our sixteen years of research on mountain goats to show how some aspects of the biology of this species play a fundamental role in its conservation. We will do so by examining the adaptations of mountain goats to their alpine environment, and by underlining differences and similarities between mountain goats and other large herbivores, in particular other mountain-dwelling ungulates.

Why Study Mountain Goats?

Mountain goats provide research challenges and opportunities from both a fundamental and an applied viewpoint. There is much concern for the conservation of this species, which appears highly sensitive to both harvest and disturbance. In addition, its alpine habitat is very sensitive to human activities and is likely at risk from the effects of climate change.

Four factors combined to provide the stimulus to study mountain goats at Caw Ridge. First, an unexplained and rapid decline in mountain goat numbers in Alberta led to the complete closure of goat hunting in the province in 1987 (Smith 1988b). Combined with the lack of information on the ecology of mountain goats, the drop in numbers convinced wildlife managers of the need for a study. Second, earlier research on mountain goats (Chadwick 1977) suggested that they may be good subjects for a study of how social behavior affects individual reproductive success and population dynamics in ungulates, because aggressive interactions seemed to be an important aspect of their ecology. Third, results obtained by studies of mountain goat population dynamics in introduced and native populations were conflicting: goats in native populations are very sensitive to even light levels of harvest (Hamel et al. 2006; Kuck 1977; Smith 1988b), while some introduced populations appear to withstand substantial harvests, similar to those normally associated with deer (Adams and Bailey 1982; Houston and Stevens 1988; Kuck 1977; Swenson 1985; Williams 1999). Finally, by providing new information about the ecology and behavior of a charismatic and economically important mountain ungulate, our work contributes to the conservation of alpine environments. Clearly, one cannot conserve mountain goats without conserving the mountain.

When our research started in 1988, what little was known about mountain goats hinted that they may differ from other ungulates in much of their behavior and ecology. Aggressive behavior was thought to be an important component of their social organization (Geist 1967, 1971), partly because their sharp stiletto-like horns (fig. 1.1) are extremely dangerous in intraspecific interactions. Females appeared to have a protracted period of maternal care, and some yearlings were reported to remain with their mothers and perhaps continue to suckle (Hutchins 1984). There were suggestions of stable associations among specific individuals, possibly indicating long-lasting relationships among female kin (Geist 1971). Therefore, mountain goats appeared to be ideal subjects for a long-term study of the effects of social status on female reproductive success, and of the relationships between behavioral ecology and population dynamics.

We set out to determine what factors affect individual differences in survival and female reproductive success and, by extension, changes in population dynamics of mountain goats. Most recent studies of large herbivores suggest that their populations are limited by a combination of forage availability and weather (Coulson et al. 2000; Gaillard et al. 2000a; Sæther 1997), although predation can also be important for populations that coexist with large predators (Owen-Smith et al. 2005; Sinclair et al. 2003; Wittmer et al. 2005). Because of its key role in population dynamics of other large herbivores (Coulson et al. 2000), we were also interested in the possible effects of changes in population density upon

Figure 1.1. Mountain goats of both sexes have black and sharp stiletto-like horns. Here, an adult female in late August. Photo by S. Côté.

individual reproductive strategies and population dynamics. There was almost no information available on age-specific survival and reproductive success in native populations of mountain goats, and nothing was known about their response to changes in population density or in resource availability. The assumption that mountain goats could be harvested at the same level as deer or bighorn sheep had led to population declines or extirpations: mountain goats may be the only North American ungulate to have suffered local extinctions through sport hunting (Glasgow et al. 2003). Weak density-dependence may explain why harvest mortality appears largely additive in this species, as suggested by a recent review of compensatory and additive hunting mortality (Lebreton 2005).

In addition to its possible significance for our understanding of individual reproductive success and population dynamics, differences in reproductive success among females may also be the key to explaining the contrasting impacts of sport hunting on mountain goat populations. Unlike most other North American ungulates, mountain goats are not obviously sexually dimorphic. To an inexperienced hunter looking through binoculars from 200 meters away, males and females do not appear very different unless they stand side by side (fig. 1.2). Most sport hunters prefer to shoot males, but only experienced hunters can identify a mountain

Figure 1.2. Sexual dimorphism is difficult to see from a distance in mountain goats. Here, the goat on the left (37 on left ear) is an adult billy surrounded by adult females. Photo by S. Côté.

goat's sex at a distance. Male goats are generally more difficult to find than females because males tend to be solitary or in small groups and spend much time in forested areas. Suppose that a hunter has approached a nursery group that includes females, juveniles, and possibly a few two-year-old males. If the selected target is the goat with the longest horns, it almost certainly will be a mature female. Now suppose that most of the recruitment within native herds is provided by mature, dominant females with long horns. The impact of removing a small number of females through sport hunting may then be far greater than that predicted by harvest models which assume that all females in a population have the same probability of contributing to recruitment. Could that be why native mountain goat populations are so sensitive to harvest? And if that is the case, which sex–age groups should be harvested, and in what proportion, if hunting is to be sustainable?

Mountain Goat Ecology and the Conservation of Mountain Ungulates

Our research on mountain goats at Caw Ridge demonstrates how long-term studies of marked individuals can both contribute to fundamental ecology and be useful for the conservation and management of ungulates.

We ask two main questions: What factors affect population dynamics of mountain goats? and, What selective pressures shape female reproductive strategy? We contend that the answers to these questions are linked by some basic biological characteristics of mountain goats that we will present in coming chapters: high and stable adult female survival, high adult male mortality and dispersal, moderate and variable juvenile survival, slow multiyear physical development, very late age of primiparity, and a strong linear and age-related dominance hierarchy among females.

Long-term studies of marked individuals provided major contributions to our understanding of the ecology of large mammals and therefore our ability to conserve them and their habitat (Gaillard et al. 2003). By monitoring marked, known-age individuals throughout their lifetime, we were able to compare different aspects of their life history, look for correlations between events that occurred in different years, and especially take into account their age. Age has a pervasive influence on almost all aspects of ungulate life history and, by extension, on population dynamics (Coulson et al. 2001; Festa-Bianchet et al. 2003; Gaillard et al. 2001). Mountain goats have a rather long life expectancy: we present here the results of our first sixteen years of work, and at the time of writing there were still two goats on Caw Ridge that had already been there when our research began. Many processes affecting population dynamics exert their effects over long periods of time: ungulate population density does not usually vary much from one year to the next, and for many comparisons (such as the effects of weather, population density, or forage quality on growth and survival) a single data point is collected each year. Consequently, a long-term study was necessary to understand the behavior and ecology of this species.

While focusing on the results of the Caw Ridge study, we will frequently compare it with the results of other long-term studies of ungulates, including other mountain ungulates such as bighorn sheep in Alberta, ibex in the Italian and French Alps, and both Alpine and Pyrenean chamois in France (Gaillard et al. 2000a). Inspired by the pioneering work on red deer on Rum, Scotland (Clutton-Brock et al. 1982), these research programs are all based on long-term monitoring of marked individuals. They have repeatedly demonstrated their value in both advancing our knowledge of population dynamics and evolutionary ecology, and in applying that knowledge to the management and conservation of wild ungulates. Unlike most other long-term studies, however, the Caw Ridge study took place in a relatively pristine environment, on a population that had not been hunted for twenty years. Although the land is scarred by resource exploration (fig. 1.3), accessible by all-terrain vehi-

Figure 1.3. A view from the eastern portion of Caw Ridge. Note the roads and trails made in the 1970s for oil, gas, and coal exploration. The trap site and cabin appear as small dots on the top right corner. Photo by J. Mainguy.

cle, and the mountain goats are occasionally harassed by helicopters and motorized vehicles, the area maintains all the wildlife species it had before European invasion, particularly the large predators such as wolves, bears, and cougars that are absent from most other long-term study areas of ungulates.

Mountain "Goat"?

The Nisga'a People of northwestern British Columbia call this white mountain-dwelling ungulate *Matx*. Europeans, however, gave it misleading names. Its scientific name means "ram of the mountain," but *Oreamnos* is not a sheep. English speakers called it "mountain goat." It is known as *chèvre de montagne* (mountain goat) in French, *cabra montesa* (mountain goat) or *cabra blanca de las Rocosas* (Rocky Mountain white goat) in Spanish and *capra delle nevi* (snow goat) in Italian. Indeed *Oreamnos* lives in the mountains and is often in the snow, so Europeans did not get it completely wrong, but it is not a goat. *Oreamnos americanus* is classified in the Tribe Rupicaprinae within the subfamily Caprinae, family Bovidae. The subfamily Caprinae includes all true mountain-dwelling ungulates, characterized by highly developed climbing skills, reliance on cliffs or steep terrain to escape predation, presence of horns in both sexes (except for some domesticated forms and one subspecies of mouflon), and for most species a complex seasonal pattern of home-range use (Shackleton 1997).

The systematics of rupicaprins are unclear: they may represent an off-shoot from other caprinae (including *Capra*, the ibex and true goats, and *Ovis*, the sheep), or they may be derived from a group ancestral to other caprins. Molecular studies disagree about the relationship between *Oreamnos* and other rupicaprins, often suggesting a phylogenetic relationship between mountain goats and muskox (Groves and Shields 1996; Hartl et al. 1990; Hassanin et al. 1998).

Rupicaprins are found in mountains from the Iberian peninsula through Europe and much of Asia to western North America. The closest contemporary relatives of mountain goats are the Asiatic serows, including the Japanese species and the much larger mainland species, at least two species of goral (*Naemorhedus* spp.), all found in Asia, and two species of chamois, which inhabit mountains from northwestern Spain to the Caucasus.

Mountain goats do not look much like true goats (fig. 1.4). They are pure white, with sharp recurved horns that resemble those of chamois and serow (fig. 1.5). The horns of males are thicker and more curved than those of females, but there is no difference in length (chapter 6). Males are also much larger than females. Sexual dimorphism increases with age, so that while male and female kids are about the same size, a full-grown adult female weighs about forty percent less than a full-grown male (chapter 6). The mountain goat is an excellent climber (fig. 1.6) and its body appears adapted to a life on the edge: the feet are short and stout, with large hooves that can open very wide, providing a strong grip on rocks and on steep terrain. Its weight is distributed vertically, which presumably helps it maintain its balance on cliff edges. From the side, mountain goats appear to have a very deep chest, but viewed from the front they are surprisingly thin (fig. 1.7).

Once people decided that this animal was a goat, they used domestic names to describe sex–age classes. We used those names in the field and we will sometimes use them here as well. A female is referred to as a "nanny," a male as a "billy," and a juvenile as a "kid."

The ancestors of mountain goats likely originated in central Asia (Geist 1971) and probably entered North America by the Beringia land bridge about forty thousand years ago (Cowan and McCrory 1970;

Figure 1.4. Mountain goats (here #137 when she was two years old; photo by S. Côté) are not true goats like ibex (an adult male from Grand Paradiso, Italy; photo by M. Festa-Bianchet) or domestic goats (purebred kiko; photo by S. Côté).

Figure 1.5. Mountain goat horns (here a two-year-old female; photo by S. Côté) resemble those of other Rupicaprinae such as chamois (photo by F. Pelletier), and Japanese serow (adult female; photo by K. Ochiai).

Figure 1.6. Mountain goats are excellent climbers. Here are some on a cliff at the west end of Caw Ridge. Photo by S. Hamel.

Rideout and Hoffmann 1975). A fossil species, *Oreamnos harringtoni*, has been found as far south as New Mexico (Jass et al. 2000). The known prehistoric distribution of *O. americanus* included Vancouver Island (Nagorsen and Keddie 2000) and possibly the Olympic peninsula. Whether or not it reached farther south than the present-day American States of Washington, Idaho, and Montana is a matter of debate, fueled by disparate interests and by preciously little data. There are no recognized subspecies of mountain goat, and little is known about its genetic variability over its geographical range.

The distribution of mountain goats includes native, reintroduced, and introduced populations (fig. 1.8). Most mountain goats are in British Columbia and Alaska (table 1.1). Including both native and introduced herds, there are somewhere between 75,000 and 115,000 mountain goats. Because of their vulnerability to hunting, mountain goats were extirpated from parts of their southern range following the arrival of European immigrants. Transplants have been used to reestablish some extirpated populations but also to introduce goats into areas with no clear evidence of their past presence as a native species. One area where mountain goats generate controversy is in the Olympic Mountains National Park in the State of Washington. Goats were introduced there in the

Figure 1.7. Viewed from the side, mountain goats appear to have a very deep chest, but viewed from the front they are surprisingly thin. A two-year-old female, #147. Photos by S. Côté.

1920s and their numbers and range greatly increased over time. Concern over their exotic status and possible negative effects on alpine vegetation led the U.S. National Park Service to adopt a policy of eradication, which was welcomed by some and denounced by others (Houston 1995; Houston and Stevens 1988; Hutchings 1995; Lyman 1988, 1994, 1995; Pfitsch and Bliss 1985). A similar situation is developing in Yellowstone National

Park, where mountain goats are not a native species but are now immigrating following an introduction north of the park in Wyoming (Lemke 2004). Mountain goats have also been introduced to the Black Hills of South Dakota and in several places in Nevada, Colorado, Montana, and Idaho (fig. 1.8), while an attempted introduction to Vancouver Island failed. In Alberta, mountain goats were reintroduced to the southwestern part of the province in 1996–1997, in areas where they disappeared in the early 1960s because of overhunting. Between 1986 and 1988, just before our study began, a few goats from Caw Ridge were captured and transplanted to southern Alberta.

Mountain goats occupy a variety of mountainous habitats, from temperate rainforest near sea level in coastal British Columbia and Alaska to xeric tundra at over 4000 meters above sea level in Colorado. Although many goat populations in the southern and western parts of their range use restricted areas with very steep cliffs, farther to the north they are

Figure 1.8. Geographic distribution of native and introduced populations of *Oreamnos americanus*.

TABLE 1.1
Estimated Numbers of Mountain Goats in North America

Jurisdiction	Year of estimate	Estimate
Alberta	2006	2750
Alaska	2000	24,000–33,500
British Columbia	2002	39,000–67,000
Colorado	2005	1965
Idaho	2000	2700
Montana	2000	2295–3045
Nevada	2006	410
Northwest Territories	2004	1000
Oregon	2006	650
South Dakota	2006	80–100
Washington	2000	4000
Wyoming	2006	270
Yukon Territory	2000	1400
Total		80,520–118,790

By separate jurisdiction. (Updated from Côté and Festa-Bianchet 2003.)

often found in rolling terrain above treeline. Mountain goats are the only caprin in much of the western half of their range, but they share parts of their eastern distribution with bighorn sheep and parts of their northern range with the two subspecies of thinhorn sheep (Dall's and Stone's). Little is known about the relationships between mountain goats and mountain sheep (either bighorn or thinhorn) in areas where both species are native, but some introduced populations of mountain goats are suspected to compete for forage or habitat with bighorn sheep and may have a negative effect on bighorn sheep populations, possibly because mountain goats are socially dominant to bighorns (Hobbs et al. 1990).

Outside national parks and other protected areas, mountain goats are hunted in most of their range. Some jurisdictions require hunters to be able to identify sex–age classes in the field, but others only specify a minimum horn length. Mountain goats do not have the same trophy popularity as wild sheep and their meat has a somewhat dubious reputation. They are an important game species in northwestern North America, where they are a symbol of mountain wilderness and are much valued economically and spiritually by Aboriginal Peoples. The Nisga'a people made clothing with the hair and hide, spoons with the horns, ceremonial regalia with the hooves, and drums with the hide. They also used the bladder for storing oil and ate the meat.

Mountain goats are possibly the least-known and least-studied North American ungulate. That is not surprising, given their northern distribution far from population centers and, especially, the difficulty of accessing much of their habitat. Most previous research on mountain goats has been based on unmarked individuals or on very small numbers of marked animals (Adams et al. 1982; Chadwick 1977; Foster and Rahs 1985; Joslin 1986; Kuck 1977; Rideout 1978; Singer and Doherty 1985; Swenson 1985). Only three studies monitored more than thirty marked individuals, and only one of those was on a native population. Smith (1986) documented survival rates of radiocollared goats in three hunted populations (two native, one introduced) in Alaska, while Bailey (1991) examined the effects of age and previous reproduction on reproductive success in large (over 120 nanny-years) samples from two introduced populations. Houston and Stevens (1988) reported the results of an experimental removal of mountain goats in the Olympic Mountains, while Hutchins (1984) studied maternal behavior in the same population.

Organization of the Book

To improve readability, we relegated the details of most statistical tests to notes at the end of some chapters, and placed some technical sections in boxes. We listed the scientific names of animals and plants in an appendix. Each chapter is preceded by a short introduction that explains its objectives and major conclusions, and the role of that chapter within the book. Each chapter ends with a summary of its primary messages.

Summary

- We will use the results of a sixteen-year study of a marked population to explore the ecology of mountain goats and our ability to conserve them and their habitat.
- Comparisons with other long-term studies of marked ungulates will underline some of the common themes of large herbivore ecology and stress some of the differences between mountain goats and other species.
- Before the Caw Ridge research program was started, there had been little research on mountain goats. In particular, there was little information based on long-term monitoring of marked individuals in native populations.
- Mountain goats are rupicaprins found in western North America,

probably related to the serow and goral of Asia and the chamois of
Europe. They live in mountains from Colorado to Alaska and occupy a diversity of habitats. Their southern distribution includes
many introduced populations in areas with no or dubious historical records of their presence as a native species.

CHAPTER 2

The Study Area and the Goat Population

Here we will introduce the study area and the mountain goat population and provide a brief history of our research. This brief history highlights the key ingredients of all successful long-term studies of wild mammals: an accessible study area (both logistically and legally), the ability to mark and monitor individuals, protection of the study population from major human interference, support from the local management agencies, stable funding, perseverance, and some luck. We then examine some of the threats to the study population from human activities, including the effects of helicopter harassment, a major and controversial current concern for the conservation of mountain goats.

The Study Area

Caw Ridge (54°N, 119°W, fig. 2.1) is approximately 30 kilometers (km) northwest of Grande Cache, Alberta, in the foothills of the Rocky Mountains. It is an ideal place to study mountain goats: it harbors the largest population in Alberta and, compared to the areas inhabited by most other goat populations, is relatively easy to access through an all-terrain vehicle (ATV) track that runs along the top of the main ridge. A test well was drilled in the 1970s near the site where we later set up our traps, but no oil was ever produced. Unfortunately, bulldozers left ugly scars on the alpine tundra of some sections of Caw Ridge. On the other hand, thanks to the old exploration roads we can reach our field camp by ATV by early or mid-June in most years, and after the snow melts we can travel the

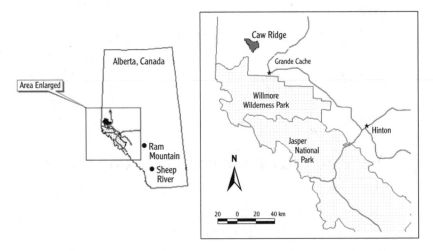

Figure 2.1. The location of the Caw Ridge mountain goat study area in Alberta, Canada. The locations of two bighorn sheep study areas (Ram Mountain and Sheep River) are also indicated.

entire length of the ridge by ATV (fig. 2.2). Most areas regularly used by goats are less than an hour's walk from the ATV trail.

Caw Ridge does not look like the typical mountain goat habitat seen in common representations of this species. Most pictures of mountain goats show them in rugged habitats, often feeding in narrow bands of vegetation above precipitous cliffs. Some of the most accessible mountain goat populations in national parks in the United States and southern Canada are indeed in areas with spectacular cliffs. Caw Ridge, however, has very few precipitous cliffs or great rock faces (fig. 2.3). It consists of a complex of rolling hills (fig. 2.4) with a few short cliff bands and several rockslides (fig. 2.5). The area used by mountain goats (approximately 28 km²) includes four major ridge complexes above timberline, all connected by well-worn goat trails (fig. 2.6). Elevations used by goats range from about 1700 meters (m) in the lowest sections of the winter range to 2180 m at the summit. Consequently, compared to many other mountain ungulate populations, the study population has limited opportunities for altitudinal migrations (Festa-Bianchet 1988d). The vegetation is mostly alpine tundra, with graminoids, forbs, and prostrate willows. Sheltered areas near creeks have extensive cover of willow bushes up to about 80 centimeters (cm) tall. Mountain avens (*Dryas* spp.) are very common throughout the ridge, but the leaves are not eaten by goats. Treeline is at

Figure 2.2. Steeve Côté shoveling snow from the quad trail to access the west end in early June 1999, a late spring. Photo by Y. Gendreau.

about 1900 m and the forest is mostly made up of spruce and alpine fir, with a few pine trees. In some places the forest ends abruptly with a well-defined treeline (fig. 2.4), in other places isolated krummholtz are dotted over the alpine tundra.

Mountain goats on Caw Ridge are geographically isolated from other goat populations. The closest large herd is at Mount Hamell, another isolated mountainous outcrop about 20 km to the southeast, inhabited by approximately eighty to one hundred mountain goats. The main Rocky Mountains range, with several other goat populations, is about 40 km to the west, but the intervening distance is almost completely covered with boreal forest and does not include many areas of goat habitat. The isolation, however, is not complete: we recorded successful immigration to and emigration from Caw Ridge (chapter 9).

The weather at Caw Ridge is typically alpine: it changes rapidly and is generally cold and often windy. Weather is a major factor affecting field-work and data collection. Snowfalls of 20 cm or more and temperatures of –4°C or less can happen at any time of the year. Summer days generally have maxima of less than 14°C. Winds strong enough to make walking or even standing difficult are not uncommon. On the other hand, there are also a few sunny days with temperatures of 18 to 20°C and a

Figure 2.3. Steeve Côté looking for kidding sites in one of the few cliffs of Caw Ridge at the west end. The rock is of conglomerate. Photo by Y. Gendreau.

light breeze. Summer days are long: in late June daylight sufficient for walking lasts from about 4:00 a.m. to 11:30 p.m. and the orange glow of the sun disappears over the horizon for only about an hour each night. Winters are long, very cold, and windy, with temperatures reaching below −40°C and only six to seven hours of daylight. The lower slopes of the ridge, however, can be warmed by Chinook winds that bring the temperature above freezing. The timing of snowmelt is extremely variable. For example, in 1998 by May 18 there were only a few patches of snow and we were able to drive ATVs to the top of the ridge. In contrast, in 1999 and in 2002, most of the ridge was still covered with 30 to 80 cm of

Figure 2.4. A view of R-3 looking east from the west end, late June 2003. Photo by S. Côté.

Figure 2.5. Big Ridge, the highest peak of the study area at 2180 m. Photo by S. Hamel.

snow in early June (fig. 2.7), and we had to shovel to get through snow drifts with ATVs until early July (fig. 2.2). The town of Grande Cache, at 1250 m elevation, has an annual precipitation of 540 millimeters (mm), including 192 cm of snow, and an average yearly temperature of 2.7°C. Precipitation on Caw Ridge, approximately 800 m higher than Grande Cache, must be substantially higher but has not been directly measured.

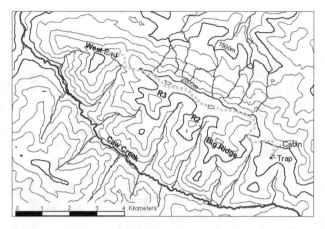

Figure 2.6. Map of the Caw Ridge study area. The dotted line is the ATV trail. Base data provided by Spatial data Warehouse, through Alberta Sustainable Resource Development.

Caw Ridge has retained all the native fauna it had before the arrival of Europeans, and in recent years may even have gained one species of ungulate, the white-tailed deer. Because of its geographical position, it combines biodiversity from the northern and the southern Canadian Rockies: eleven species of large mammals can be found in the area. The only other ungulate regularly found at the elevations used by goats is the bighorn sheep, which is near the northern limit of its geographical range. A large herd of bighorn sheep uses areas just east of the section of Caw Ridge frequented by mountain goats. Although interspecific competition was not one of the subjects we investigated, it was obvious that there was a sharp spatial segregation between the two species, with an imaginary boundary line at the trap site (fig. 2.6). Sections of Caw Ridge east and west of the traps did not look very different to us, but in most years it was unusual to see a bighorn sheep west of the traps, or a mountain goat east of the traps. Both species came to the traps, although bighorn sheep visited them much more rarely than mountain goats and later in the summer. Part of the reason for the spatial separation of the two species may be an open-pit coal mine situated approximately 4 km east of the traps (fig. 2.8). Mountain goats are very sensitive to human disturbance and may have avoided the area close to the mine that was also the main sheep habitat. Bighorn sheep thrive in reclaimed coal mines (MacCallum and Geist 1992), but goats do not appear nearly as capable as sheep to habituate to human activities. We saw few interactions between sheep and goats, and

Figure 2.7. Caw Ridge in early June 2002, a very late spring. The box traps were buried in snow. Photos by S. Côté.

usually the goats vigorously drove away the sheep, although in a few instances they ignored them. At treeline and at lower elevations on Caw Ridge one finds moose, wapiti, mule deer, and white-tailed deer. All of these species were occasionally seen in areas used by mountain goats. A herd of about 300 to 350 woodland caribou (Edmonds 1988) migrates through Caw Ridge, making substantial use of the alpine tundra. The caribou move from east to west in April–May, on their way to calving

Figure 2.8. The coal mine about 4 km east of the cabin. Photo by J. Mainguy.

grounds and summer ranges in the mountains, and repeat the journey from west to east in October–November, returning to wintering areas in the boreal forest. Caribou winter ranges north and east of Caw Ridge are under increasing pressure from logging activities and oil and gas exploration (Smith 2004; Smith et al. 2000), and the future of woodland caribou in Canada is uncertain (Thomas and Grey 2002).

We documented three species of carnivores on Caw Ridge as predators of mountain goats (Festa-Bianchet et al. 1994): grizzly bears, wolves, and cougars. Other potential goat predators include black bears, wolverines, coyotes, and golden eagles. Golden eagles were seen during most days while we conducted fieldwork in summer, possibly because of the abundance of smaller mammals. Rodents and lagomorphs commonly seen in the alpine tundra include hoary marmots, pikas, Columbian ground squirrels, and golden-mantled ground squirrels. Bushy-tailed woodrats and deer mice are also common but are less frequently seen because of their nocturnal habits. Porcupine and snowshoe hares are sometimes seen near treeline. Although grizzly bears are of greatest concern for personal safety, the greatest material damage was inflicted by hoary

Figure 2.9. Hoary marmots damaged field equipment, especially quad parts with perspiration. Photo by E. Cardinal.

marmots, who gnawed everything imbibed with perspiration: the seats and handles of ATVs, the inside of helmets, backpacks, coolers, cables for the platform scales, and so on (fig. 2.9). The scientific names of all mammals and birds that we saw on Caw Ridge are in the appendix.

Recent History of the Caw Ridge Mountain Goat Population

Until about 1970, Caw Ridge could only be reached on foot or horseback, and until a few years earlier, one could not drive to within less than about 60 km. The town of Grande Cache was established in 1969, when oil, gas, and coal exploration activities commenced and bulldozer tracks made their first appearance. Until 1969, hunters reached the area on horseback after a few days of riding. A dry-weather road was built in the early 1970s to allow access to a test oil well on the site where later we installed our traps. When the Willmore Wilderness Park was first established in 1959, it did not include Caw Ridge, but it did include nearby Mount Hamell, which is inhabited by another goat population, as well as the current site of the town of Grande Cache. The park lies immediately north of Jasper National Park. Resource extraction, forestry, and motorized vehicle access are prohibited within Willmore Park. The park's boundaries were changed in 1962 to exclude coal-rich areas, exemplifying how the Alberta government strikes a balance between biodiversity and the resource-extraction industry to this day. Hunting of mountain goats on Caw Ridge was allowed until 1969 under a general season with a limit of one goat per hunter, but was closed in 1970. Reports from the mid-1970s suggest that the goat population at that time totaled only fifty-five individuals (McFetridge 1977), but it is unclear whether that estimate reflected a low population because of excessive hunting until a few years earlier or low censusing efficiency. Aerial surveys between 1979 and 1986 consistently reported about 80 to 90 goats (Smith 1988b), but given the efficiency of aerial surveys on Caw Ridge (Gonzalez-Voyer et al. 2001), the true population in the 1980s was likely between 110 and 130 goats, similar to or slightly greater than the levels we recorded during the first half of our study. Although the potential impacts of hunting until 1969 are unknown, when our study began the population had not been legally hunted for twenty years, or almost three goat generations. Poaching of mountain goats in Alberta, and on Caw Ridge in particular, appears to be minimal. Although by definition one cannot easily measure the extent of an illegal and covert activity, we have no evidence of poaching of mountain goats on Caw Ridge during our study.

Despite protection from hunting, however, the Caw Ridge mountain goats have experienced artificial manipulations of their numbers. Twenty goats, including nine adult females, were removed for relocation to southern Alberta or the Calgary Zoo in 1986–1988, and seven died during capture operations for our research program (table 2.1). In addition, one adult female broke a leg in 1995 after being harassed by a resource-exploration helicopter (Côté 1996) and as many as seven kids were abandoned by their mothers following capture (Côté et al. 1998a).

Before our work began, there had been no study of mountain goats on Caw Ridge, other than a visit by McFetridge (1977) in August 1975 when he counted fifty-three goats. The Alberta Fish and Wildlife Division, however, conducted helicopter censuses of mountain goats on Caw Ridge since the mid-1970s. The accuracy of those surveys is discussed in chapter 3. The transplant program in 1986–1988 mostly targeted non-lactating nannies, partly to avoid leaving behind orphaned kids. In 1987 and 1988, a few goats were caught that were deemed unsuitable for relocation: they were marked and released, and constituted the beginning of our marking program. When our research began in June 1989, there were already twelve marked goats in the population. The numbers of goats in the study population increased during most years of the study and the proportion marked also increased (table 2.2; fig. 2.10).

TABLE 2.1
**Fate of Mountain Goats Artificially Removed from
Caw Ridge in 1986–2003**

Year	Goats removed	Why
1986	3 adult females	Transplanted to Highwood Range and Calgary Zoo
	3 yearling females	
	1 yearling male	
	1 yearling female	
1987	1 yearling male	Transplanted to Livingstone Range
	4 adult males	
	1 2-year-old female	
	3 3-year-old females	
	1 adult female	
1988	1 adult female	Transplanted to Livingstone Range
	1 adult male	
1990	1 female kid	Capture mortality
1991	1 female kid	Killed by trap door
1993	1 4-year-old female	Capture mortality
	1 yearling male	Capture mortality
1995	1 3-year-old female	Capture mortality
2000	1 2-year-old male	Capture mortality
	1 male kid	Killed in trap by another goat

A Brief History of the Caw Ridge Mountain Goat Study

The idea of studying the Caw Ridge goats developed during a conversation between Marco Festa-Bianchet and Kirby Smith in April 1988. Kirby was the wildlife manager responsible for this area. He had used a drop-net to capture goats on Caw Ridge and was interested in a study of population dynamics. A recent drop in mountain goat numbers in Alberta had led first to a drastic reduction in the number of hunting licenses sold, and eventually to the closure of hunting over the entire province. Caw Ridge was clearly the best population in Alberta for a study based on monitoring marked individuals. Later that year, Marco went to Caw Ridge for the first time and a few goats were caught, marked, and released.

The study began officially in June 1989, while Marco was a postdoctoral research associate at the Large Animal Research Group of the University of Cambridge, England. Martin Urquhart was hired as a field assistant and did most of the fieldwork during the first five years of the

TABLE 2.2.

Mountain Goats in the Caw Ridge Study Area from 1989 to 2003

Sex–age class	Year														
	1989	1990	1991	1992	1993	1994	1995	1996	1997	1998	1999	2000	2001	2002	2003
Adult males	20	16	17	19	20	12	18	18	19	22	26	36	31	39	36
	(5)	(13)	(14)	(16)	(19)	(12)	(18)	(18)	(18)	(19)	(21)	(31)	(28)	(36)	(35)
Adult females	41	35	40	41	44	45	53	54	52	51	47	50	54	59	58
	(14)	(23)	(28)	(34)	(40)	(44)	(51)	(52)	(50)	(50)	(46)	(48)	(54)	(59)	(58)
Yearlings	14	12	13	8	13	20	15	14	8	13	33	17	21	13	25
	(10)	(10)	(8)	(4)	(10)	(14)	(12)	(10)	(7)	(11)	(13)	(10)	(16)	(8)	(24)
Kids	21	18	19	17	26	28	28	21	23	36	29	35	18	36	33
	(13)	(11)	(4)	(10)	(10)	(5)	(9)	(7)	(11)	(0)	(0)	(0)	(2)	(0)	(0)
Total	96	81	89	85	103	105	114	107	102	122	135	138	124	147	152
	(42)	(57)	(54)	(64)	(79)	(75)	(90)	(87)	(86)	(80)	(80)	(89)	(100)	(103)	(117)

Numbers refer to the June population, including all kids born each year, even though a few kids were born in July or were seen in May but died before June. The numbers in parentheses indicate marked individuals by the end of each summer. The count for 1989 indicates the number of goats in August but includes one kid known to have died earlier that summer.

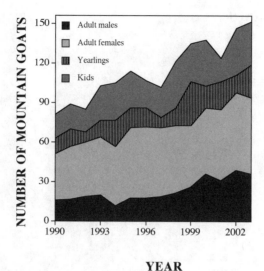

YEAR

Figure 2.10. Changes in numbers and sex–age structure of the Caw Ridge mountain goat population in June, 1990 to 2003.

study. For the first two years, fieldwork continued through the year, but since 1991 the field season has been limited to mid-May to late September or early October. Work on Caw Ridge during winter is costly and dangerous because of the remoteness of the study area, short days, and harsh weather. From 1989 to 1991 the field crew lived in a fiberglass cupola (fig. 2.11) salvaged from a forestry fire lookout. Two years later we assembled a plywood cabin (fig. 2.12). Living conditions and our ability to collect field data improved tremendously as a result.

In 1990 Marco was hired by the Department of Biology of the Université de Sherbrooke in Québec, where scientific productivity was valued. Employment as a professor meant the implication of graduate students in the mountain goat research. Long-term financial support from the Natural Sciences and Engineering Research Council of Canada (NSERC) brought financial stability to the program. Steeve Côté became involved in 1994 as a graduate student. He graduated in 1999 (Côté 1999) and took over direction of the research in 2001 after obtaining a faculty position at Université Laval in Québec City. The study continues thanks to the involvement of graduate students, funding from various agencies, and the cooperation and support of the Alberta Fish and Wildlife Division.

Figure 2.11. The fiberglass lookout cupola used as living quarters from 1989 to 1991. It is now used for storage. Photo by S. Côté.

Figure 2.12. The cabin built in 1991 with the help of inmates from the Grande Cache correctional center and used as our field base ever since. Photo by S. Côté.

Threats to the Caw Ridge Mountain Goats

The Caw Ridge mountain goats were mostly undisturbed by humans and lived in a relatively pristine environment until about 1970. Today, however, they find themselves at the center of intense and encroaching human activities. The area is easily reached by ATV and only landslides and deep snow currently restrict motorized access. ATV enthusiasts arrive on Caw Ridge on most weekends from late June to September. Most visitors remain on the main trail and their impact is minimal, but there have been instances of individuals chasing goats with ATVs, or driving over fragile alpine tundra. In addition, motorcycles riding off trail are a common sight in summer. Snowmobile access is difficult but possible from November to May. Only the relative isolation of Caw Ridge has protected it from the kind of ATV abuse that characterizes many other areas of Alberta: it takes more than two hours by truck and ATV to reach the ridge from Grande Cache, and the closest major population center (Edmonton, a city of about a million people) is over five hours' drive from Grande Cache. Caw Ridge is publicized by Alberta's "Watchable Wildlife" program as the best place in the province to observe mountain goats. Overall, the current level of recreational use does not appear to have strong negative effects on the goats, but the number of ATVs increases every year (from about fifty vehicles per summer in 1994 to over three hundred now) and may soon reach a point where it will have to be controlled.

A much more serious threat is represented by the resource extraction industry (Côté 1996; Joslin 1986). Over twenty-five years ago, McFetridge (1977) pointed out that plans were under way to mine coal on Caw Ridge, which he already recognized as one of the most important mountain goat habitats in Alberta. Coal mining began approximately 20 km southeast of Caw Ridge in the late 1960s and expanded to within 4 km of the trap site in 1996. Exploration activities took place in the summers of 1997 and 1998 on the eastern part of the ridge. Those activities involved drilling test holes twenty-four hours a day and bulldozing a fish-scale pattern of new scars on the fragile alpine tundra. Although access to the exploration sites by trucks and bulldozers was mainly from the east (therefore not over areas used by mountain goats) and exploration activity was located approximately 2 km from the closest area regularly used by the goats, all areas within sight of the activity (including the trap site) were temporarily abandoned by mountain goats while exploration crews were working, confirming earlier suggestions (Côté 1996; Joslin 1986) that mountain goats are extremely sensitive to human disturbance. Mountain goat behavior is in marked contrast to that of bighorn sheep, which tolerate industrial activity and readily use reclaimed open-pit coal mines (MacCallum and Geist 1992). The closest pit where coal is mined is now just over 4 km from the traps. From the eastern end of Caw Ridge to the west, the view is of relatively pristine alpine tundra, scarred by a few bulldozer tracks. To the east there is a giant black hole (fig. 2.8). Perhaps in the future the view to the west will also be that of an open-pit coal mine devoid of mountain goats.

Mountain Goats and Helicopters

A major early contribution of the Caw Ridge study was the injection of scientific data into the debate on wildlife harassment. That injection took place in the form of a quantitative study of how mountain goats react to helicopters flying over their habitat (Côté 1996). Exploration companies in remote parts of Canada use helicopters to supply work camps and support crews looking for oil and natural gas. Until 1996, many helicopters routinely flew over Caw Ridge.

The effect of helicopters on mountain goats had long been the object of speculation, with environmentalists and most biologists suggesting that helicopter use was highly detrimental to mountain goats, and resource companies and promoters of helicopter-based recreation (a rapidly expanding industry) arguing that the impact of helicopters was minimal. In the absence of any real information, neither side provided much evidence supporting its view. That is unfortunately typical of many situations involving wildlife harassment: data are scarce or inconclusive, and people argue over anecdotes. Preventing wildlife harassment inevitably requires curtailment of human activities, often at considerable cost for those involved. To justify intervention and modification of human behavior, it is therefore important to clearly demonstrate that harassment has negative effects.

Initially, our reaction to helicopters flying over mountain goats was frustration. The typical consequence of a flight was that observations were compromised or ended for that day, since the goats typically fled the helicopter until they ran out of sight, or behaved nervously for several hours. To put it plainly, if a helicopter flew over goats that we were observing our day was shot. We usually reported the incident, and the helicopter operator was made aware of our complaint, but there were no sanctions and within a few days the episode would repeat itself. In 1995 we began systematically recording the reactions of mountain goats to helicopters. By the end of the summer, many episodes of helicopters flying within sight of mountain goats were available for statistical analysis.

TABLE 2.3
Reaction of Mountain Goats to Helicopters Flying at Different Distances, June–August 1995

	Mountain goat reaction (number of groups)		
Distance (m)	Light	Moderate	Strong
< 500	0	3	17
500–1500	0	3	4
> 1500	34	15	5

See text for definition of behavioral reactions. (From Côté 1996.)

Not surprisingly, the goats' reaction to helicopters varied according to the distance from the aircraft (table 2.3). Even when the helicopter was over 1500 m away, however, the goats reacted strongly in 9% of cases. Helicopters flying within less than 500 m elicited a strong reaction 85% of the time. A "strong" reaction meant walking or running for over 100 m, or staying alert for over ten minutes, while a "moderate" reaction involved a movement of 10 to 100 m and two to ten minutes spent alert (Côté 1996). We documented groups being broken up, kids temporarily separated from their mothers, one female breaking a leg while fleeing a helicopter, and several cases of panicked goats running at full speed over cliffs and precipitous terrain. When helicopters flew within 2 km, the goats were always disturbed, although in some cases they only moved a short distance or were alert for less than two minutes. The goats have not habituated to helicopter flights and data from recent years reveal the same patterns.

Our research confirmed that mountain goats are more sensitive to disturbance by helicopters than other ungulates, as was suspected by many biologists. Bighorn sheep usually do not react to helicopters over 500 m away (Stockwell et al. 1991), and a minimum distance of 1 km has been recommended to prevent helicopter harassment of caribou and muskox (Miller and Gunn 1979). The behavior of goats subjected to helicopter harassment corroborates the suggestion that this species is highly sensitive to several kinds of human disturbances, including scientific research (chapter 3). The greatest value of the helicopter study, however, was the application of systematic data collection and analysis to a serious conservation problem, leading to publication in a scientific journal widely read by North American wildlife managers. Before this study was published, managers had based recommendations about helicopter use over mountain goat habitat on anecdotes. Anecdotes have little impact on policy decisions opposed by powerful economic interests, and counter-anecdotes can be readily produced: helicopter operators would simply say that mountain goats were not disturbed by helicopters.

We recommended that helicopters should not fly within 2 km of mountain goat habitat. That recommendation was welcomed by wildlife managers but met with hostility from some helicopter tour operators. Helicopter harassment of wildlife over alpine areas is not a trivial issue. Both industrial and recreational uses of helicopters are increasing. Heli-hiking, heli-skiing, heli-fishing, heli-biking, heli-barbequeing, as well as helicopter sightseeing tours, are rapidly expanding in the Canadian Rockies and in other mountainous areas in North America and Europe. An article in a Canadian magazine even praised helicopter-based tourism as a family activity. Unfortunately, little work has been carried out on alpine wildlife exposed to harassment from increasingly frequent flights. Our work on mountain goats was one of the earliest alarm bells.

What are the potential consequences of helicopter harassment? The broken leg suffered by one nanny suggests that immediate consequences can be serious. Goats often left areas where they had been disturbed. If helicopters flew repeatedly over certain areas (as is the case in sightseeing tours and flights to provide recreational access), we expect that goats will abandon those areas, thus reducing the available habitat. Forced to con-centrate in smaller areas, mountain goats will have access to a smaller quantity of food and may be more vulnerable to predation (chapter 4). Clearly, less available goat habitat will mean fewer goats. In addition, the mother–kid separations caused by low-flying helicopters increase the risk of predation on the kid while away from its mother and may lead to per-manent separation, given that lactating nannies do not seem to have a very strong maternal instinct in stressful situations (chapter 3). Intense helicopter harassment could lead to the decline of mountain goat popu-lations. Recent work on chamois suggests that other forms of aerial ha-rassment, such as paragliders, may also affect the behavior and area-use patterns of mountain ungulates (Enggist-Düblin and Ingold 2003). Ha-rassment by helicopters and other aircraft is a serious threat to the con-servation of alpine ungulates.

Whether or not mountain goats may habituate to helicopters remains an open question. It is not impossible that goats may learn to tolerate hel-icopters flying within 2 km. In some national parks, mountain goats have become more tolerant of humans and vehicles, following decades of ex-posure and very few instances of negative conditioning. Our study area, however, has been at the center of intensive industrial exploration activi-ties for over thirty years and the goats show no habituation. We recommended a 2-km buffer based upon observation of behavior of goats that had been exposed to helicopter flights over their entire lives. The results of future quantitative studies may be different from ours, but data cannot be countered with anecdotes. Our recommendation of a 2-km buffer zone stands. Those who choose to spend money on helicopter-based recreation should be aware of the negative effects of their choice on alpine wildlife.

Summary

- Research on mountain goats at Caw Ridge began in 1989, although a few goats were marked and released in 1987 and 1988.
- The study area was selected because of its accessibility and large mountain goat population, unhunted since 1969. It retains all species of large mammals present at the time of European contact, including several carnivores that prey on goats of all sex–age classes.
- Coal mining is a threat to mountain goat populations, including the study population.
- Helicopter harassment is a serious problem for the conservation of mountain goats. Helicopters should not fly within 2 km of mountain goat habitat.

CHAPTER 3

Caw Ridge Study Methods and Limitations

This chapter describes the main techniques used to collect data during our research on mountain goats. Procedures specific to a particular type of data are described in greater detail in the appropriate chapters. The central technique of our research is simple: we caught mountain goats as young as possible, marked them, then monitored them over their entire lives. Here we also point out some of the logistic limitations of our research, including those caused by the unexpectedly high susceptibility of our study species to human disturbance and those due to the inaccessibility of the study area during much of the year.

The Field Season

Fieldwork typically lasted from mid-May to September. During the first two years of the study, fieldwork continued during winter. Especially during the winter 1990–1991, however, it became evident that work at Caw Ridge at that time was very difficult because of short days and harsh weather. Therefore, since 1991 the field season has been limited to late spring through to late summer or early fall. We usually arrived before kids were born and spent the first two weeks searching for parturient nannies. At the beginning of June, our activities settled into a pattern that continued until September. In the morning, the traps were checked every 30 to 40 minutes beginning at first light, which varied from 4:30 to 6:00 a.m. as the summer progressed. If no goats were trapped, the crew closed the traps at about 8 or 9 a.m. and left on ATVs to search for goats and spend the day making behavioral observations. In the evening the traps were reopened, we prepared a meal, then connected the cellular phone to a battery for two hours of potential communication with the outside world while doing vast amounts of paperwork to record the day's observations.

Catching, Measuring, and Tagging Goats

Most of our research objectives required long-term monitoring of marked individuals. Our basic technique was to mark each goat so that we could identify it during observations, then monitor aspects of the ecology (survival, reproduction, area use) and behavior (foraging, social interactions, maternal behavior) of that individual for its remaining life span. While still somewhat foreign to North American wildlife managers, this approach has amply demonstrated its superiority for ecological and behavioral studies of large mammals (Byers 1997; Clutton-Brock et al. 1982; Festa-Bianchet et al. 1998; Gaillard et al. 1997; Hogg and Forbes 1997; Lunn et al. 1994). Recent research has shown that for long-lived animals such as ungulates with very strong effects of sex and age on survival and reproduction, there is little hope of understanding either the evolution of reproductive strategies or the processes affecting population dynamics without long-term studies (Coulson et al. 2001; Gaillard et al. 2000a; Gaillard et al. 2001).

In 1989, we installed two box traps and four Clover traps (Clover 1956; fig. 3.1) at the site where goats had been previously baited to a drop

Figure 3.1. The trap site showing Stevenson's wooden box traps and self-tripping Clover traps of nylon netting over a metal frame. The picture is taken from the blind from where we pull the ropes to close the trap doors. Photo by S. Hamel.

net. We used blocks of salt for bait. The goats were initially attracted to this site by an unknown greasy substance left behind after a test well was removed in 1976. Therefore, when we began our capture program the goats were already coming to the trap site to lick the ground. Over the years, we increased the number of traps to eight box traps and six Clover traps. Between 1995 and 1998 we installed two additional Clover traps at the western end of Caw Ridge, mainly to increase the number of captures of adult males that seldom visited our main trap site. We never set traps in other locations, even though that may have increased the number of captures. Goats had to be processed immediately upon capture to avoid problems with kid abandonment or with injuries resulting when two goats were accidentally caught in the same trap. Therefore, traps were set only where we could regularly and closely monitor them. In recent years, we stopped using the Clover traps and now only use remote-controlled box traps, which allow us to choose which goats to catch.

We caught and processed as many as seventeen goats in one day, but in most cases we only processed one to three individuals in a day. Sometimes the goats would return to the traps two or three times in the same day, especially in mid to late June, but other times we would go for weeks without any captures. The frequency of goat visitations to the traps declined in August and September, so that most captures occurred in June and July.

Mountain goats are very aggressive when confined in a trap, and use their horns to defend themselves. Captured goats tried to stab us whenever we were within reach. Steeve once received a deep puncture wound on a leg after approaching an adult male caught in a Clover trap with loose nylon netting. Before processing, most captured goats two years of age and older were immobilized by intramuscular injection of xylazine hydrochloride (Rompun®) which was later reversed by intramuscular injection of idazoxan (Haviernick et al. 1998). We handled all kids and most yearlings without drugging them.

Kids were marked with small Allflex plastic ear tags in one or both ears with unique color combinations (fig. 3.2). Between 1988 and 1995, thirty-one kids also received drop-off radio collars (table 3.1). Yearlings of both sexes and adult males were marked with unique color–number combinations of Allflex ear tags (fig. 3.2). Most adult females were marked with a combination of ear tags and visual collars (fig. 3.2). One unmarked adult female was identified by her physical appearance between 1994 and 1999, when she was the only unmarked nanny older than three years. Some goats also received VHF radio collars (fig. 3.2; table 3.1). Radio collars were used to find carcasses to determine cause of death

Figure 3.2. Markings to identify individual goats at a distance (a) adult female #63 with large ear tags and kid #180 with small ear tags (photo by S. Côté), (b) yearling #304 with medium tags (photo by M. Houle), (c) adult female #183 with a visual collar (photo by S. Hamel), and (d) adult male #249 with large ear tags and a VHF radio collar (photo by S. Hamel).

TABLE 3.1

Minimum Number of Mountain Goats with Functioning Radio Collars on Caw Ridge, 1989–2003

Year			Sex–age class		
	Kid	Yearling	Adult male	Adult female	Total
1989	9	2	0	2	13
1990	10	4	3	11	28
1991	2	5	3	10	20
1992	4	1	2	11	18
1993	3	1	2	10	16
1994	0	3	1	10	14
1995	1	0	1	12	14
1996	0	0	5	9	14
1997	0	0	9	6	15
1998	0	0	7	4	11
1999	0	0	8	3	11
2000	0	0	7	2	9
2001	0	0	7	1	8
2002	0	0	7	0	7
2003	0	0	12	0	12

The list includes all goats whose collar provided at least one radio location in a given year.

(particularly of kids between 1989 and 1993), to monitor potential emigrants and to help find groups of goats, particularly in forested areas. Most collars had a mortality switch, so that the signal rate doubled after a few hours of complete inactivity.

At capture, each goat was weighed with a spring scale to the nearest 0.5 kg, and body and horn measurements were collected with a measuring tape. Lactation status of nannies was determined by udder examination. A hair or tissue sample was collected for genetic analysis from all individuals captured or recovered dead from 1995 onward.

Remote Sensing for Mountain Goats

Practical and ethical reasons limited our capture program but methodological advances allowed us to obtain information on individual goats without having to capture them. Although information on genetics and pedigrees from molecular analyses will not be presented here, improved DNA amplification techniques now provide genotype information from hair and fecal samples. In addition, in 2001 we installed an electronic

BOX 3.1
Problems Handling Goats

Initially, we expected that captures would have minimal effects on mountain goats. Our expectation was based upon successful capture programs of bighorn sheep, where hundreds of sheep had been caught in a corral trap (Festa-Bianchet et al. 1996) or darted free-ranging (Jorgenson et al. 1990) with very few negative consequences, and published reports of goat captures that did not mention negative effects (Bailey 1991; Houston et al. 1989; Smith 1986). We soon discovered, however, that just as they are more sensitive to harassment than other ungulates, mountain goats are also more sensitive to being handled. We discontinued capture practices that had unwanted effects on the goats because of ethical concerns and because we did not want our results to be a function of our methods.

The first problem we encountered was kid abandonment by the mother following capture, first noticed in 1992. We determined that the risk of abandonment was higher if the kid's mother had been caught and drugged (five of thirty-one kids) than if the mother had not been handled (one of forty-six kids, where the mother was not captured or was released without being handled) (Côté et al. 1998a). Consequently, beginning in 1995 we did not handle any lactating nanny. When we captured nanny–kid pairs in the same box trap, we caught and removed the kid, measured and tagged it, then opened the trap door fully and released the kid and its mother together. We also did not capture any kid before mid-July. Despite the new procedure, we experienced problems in four of thirty-four kid captures between 1995 and 1997, mostly involving kids that were not seen again after capture and therefore may have died as a result of the capture. Therefore, beginning in 1998 we did not handle any kids or lactating nannies. Consequently, we were unable to determine the maternity of many goats born after 1997, with the exception of those that remained associated with their mothers as yearlings and were caught while the association was still evident (chapter 5).

We discovered two other problems related to our capture and marking operations: kids with radio collars appeared less likely to survive than kids with ear tags during years with below average kid survival, and chemical immobilization of young nannies delayed primiparity. In 1989 and 1991–1993, 60% of fifteen kids with radio collars survived, compared to 82% of seventeen kids without radio collars ($G = 1.99$, $P = 0.16$; Côté et al. 1998a). Although the difference in survival was not significant, it led us to stop putting radio collars on kids. The negative effects of drugging on reproduction by young females were as clear as they were unexpected. Of twenty females aged three or four years that were caught and drugged, only six (30%) produced a kid the following year, half the rate (61%) of forty-four females of the same age that had not been drugged (Côté et al. 1998a). Therefore, we stopped catching females

BOX 3.1
(Continued)

aged three or four years. We did not detect any effect of drugging on reproduction of older females (Côté et al. 1998a).

Stunned by these unexpected and depressing results, we examined our data
to look for other potential negative effects of capturing goats but found none.
Radio collars had no effect on survival of older goats, and for those kids that
were not abandoned, ear tags did not affect survival. As reviewed in Côté et al.
(1998a), most studies of ungulates that examined the effects of handling and
drugging, or the effects of artificial markings, found at least some undesired
consequences. In most cases, however, the effects were less drastic than those
we found, further reinforcing our impression that mountain goats are more
likely than other ungulates to suffer adverse effects from human activities. Because of the growing and justified concern for ethical treatment of research
animals, it behooves wildlife researchers to ensure that field techniques do not
affect the behavior (Pelletier et al. 2004), survival, or reproduction of study animals. A common problem is the lack of control samples: because unmarked
animals cannot be monitored, it is difficult to test the effects of marking. Our
results and those of other studies, however, imply that it is clearly not acceptable to simply assume that captures, marking, and other manipulations have
no effects.

platform scale with a digital remote display and weighed goats by luring
them onto the platform with a block of salt (fig. 3.3). Because of the aggressiveness of mountain goats, usually only one goat stepped on the
scale at a time, and subordinates waited their turn while dominants licked
the salt and were weighed. Kids were mostly weighed when they stepped
on the scale with their mother. This system for weighing free-ranging
ungulates was first used for Alpine ibex and we have since used it for
bighorn sheep (Bassano et al. 2003). Over the next two years we added
two platform scales, so that more goats can be weighed when they visit
the trap site.

Censusing Mountain Goats and Monitoring
Reproductive Status

Weather permitting, we searched for mountain goats on most days during the field season, locating groups either visually or from radio signals.
When a group was sighted, it was usually approached to within 200 to

Figure 3.3. Platform scale installed between box traps to weigh free-ranging goats. Photo by S. Hamel.

700 m to identify individuals with spotting scopes. We noted the ID number of each marked goat. Unmarked goats were classified as kids, yearlings, two year olds, or adults. All age classes except kids were also classified by sex using horn morphology (Smith 1988a). We determined the sex of most unmarked kids by observing their urination posture (females squat and males stretch), or the external appearance of the anogenital patch during suckles, when kids hold their tail vertical (fig. 3.4). For females, we also noted at each sighting whether they were associated with a kid, a yearling, or a two year old. We considered suckling, body contact while lying, or close and persistent following as evidence of mother–offspring association. All yearlings or two year olds that had been marked as kids and whose mother was known only associated with their mother. We also noted the Universal Transverse Mercator coordinates of each group of mountain goats to the nearest 100 m. With the exception of some adult males, most marked goats were seen between fifteen and fifty times each year; therefore, we had accurate information on area use, survival, and reproductive success. Males made extensive use of forested areas (chapter 4) and were often solitary or in very small groups. Some adult males were only seen four or five times during a season, especially in the earlier years of the study. We assumed that any goat not seen

Figure 3.4. Kids were sexed by looking at the external appearance of the anogenital patch during suckles, when kids hold their tail vertical. Note the anogenital patch of a female kid. Photo by S. Côté.

during a year had either died or permanently left the study area. Three adult males, however, left for one to four years then returned to Caw Ridge.

Population estimates for the first four years of the study were based on the highest single-day classified count each year, or on a combination of the highest counts of "nursery" goats and of adult males, usually obtained on different days. On the days when these highest counts were obtained, we saw all marked goats. From 1993 onward, the few unmarked goats were recognizable, and population estimates were the total counts of individually recognizable goats.

By intensively searching the study area in late May and early June, we determined birthdates of most kids to within three days. In many cases, we either witnessed the birth or the kid's appearance and behavior suggested that it was less than two days old. In 1989, we caught two females that showed evidence of lactation but were not seen nursing kids: we assumed their kids had died. All other females classified as lactating were seen nursing kids. Because of the long hair of mountain goats in early summer, it was not possible to determine reproductive status by noting the degree of swelling of the udder during field observations. Therefore, we may have missed cases where kids were born but died at or soon after birth, or cases of twinning where one sibling died before we saw it. We suspect that there were very few cases when we missed neonatal deaths. Parturient females normally isolate themselves a few days before giving birth and remain alone with their kid (or their kid and their yearling), for at least five days after giving birth. Since 1994, we saw only two females that left the nursery groups during the parturition season, but that were not subsequently seen with a kid. Those two females probably had kids that died at or soon after birth. All other females that we classified as barren since 1994 (228 female-years) remained within nursery groups during the parturition season, reinforcing our impression that they did not give birth.

After a kid was marked, its mother was identified by behavioral association. Weaning success was defined as survival of the kid to early September and was measured for all marked nannies each year, regardless of whether or not their kids were marked. Survival to one year, however, could only be determined if the kid was marked or if the surviving yearling associated with its mother. Not all yearlings associated with their mother through the summer (chapter 5), but most remained with their mother until just before the parturition season. When a nanny weaned an unmarked kid in September but was not accompanied by a yearling in late May the following year, however, we could not know if her kid had died during the winter or if it was one of the few unmarked yearlings that were alive but no longer associating with their mothers. Therefore we do not have complete data on survival of kids to one year of age for all nannies.

Observing Goat Behavior

Beginning in 1991, we made three types of behavioral observations: focal sampling of marked individuals, scan sampling, and all-occurrence sampling of social interactions (Altmann 1974). Focal samples lasted from five minutes to several hours depending on the type of behavior studied, and scan sampling could last for several hours. Both techniques are explained in greater detail in the appropriate chapters. For social interactions, we attempted to note the identities of the initiator and receiver, the behaviors used, the winner and loser of the encounter, and whether the winner replaced the loser at its foraging or bedding site (Côté 2000). Although fewer than 1% of interactions involved contact, most were not subtle and it was easy to identify winner and loser. The loser displayed submissive behaviors and almost always moved away (Côté 2000).

Fecal and Vegetation Samples

We collected fresh fecal samples from ten to twelve individuals at bi-weekly intervals from late May to late August each year. Samples were air dried and later analyzed for nitrogen content by micro-Kjeldahl digestion. In 1991, we installed fourteen 1-m² wire mesh exclosures on a grassy slope at the western end of Caw Ridge that was used year-round by the goats (fig. 3.5). We clipped a 20 × 20 cm quadrat inside and outside each exclosure in the first weeks of June and September in most years. After collecting the sample, the exclosure was moved by 1 or 2 m and secured to the ground. Quadrats were clipped of all vegetation to 1 cm from the ground, except for the leaves of mountain avens (*Dryas* spp.), which were abundant but apparently not eaten by mountain goats. Vegetation samples were air dried and later oven dried and weighed.

Limits of the Caw Ridge Study

Although Caw Ridge is one of the best possible study areas for mountain goats, the scope of our research was limited by technical problems. The greatest disadvantage of Caw Ridge is the harsh weather and difficult access from November to May. Reaching our cabin in mid-May often took

Figure 3.5. Steeve Côté by a 1-m² wire mesh exclosure installed to measure vegetation growth at the west end, late September 1994. Photo by K. Seidel.

more than a day of effort! What happened during the eight months of winter is essentially a black box to us, with the exception of the observations made over the first two years of the study. Because we did not observe the rut until 2004, we have no information on male reproductive behavior. However, concomitantly with a genetic study, we are now observing goats during the rut. Our information on the behavior of adult males once they left the nursery groups is much less extensive than for other sex–age classes, mostly because males made extensive use of forested areas (chapter 4) and were seldom observed for long periods of time.

We have limited information on body and horn measurements. At the start of the study we hoped to recapture and remeasure most individuals every year, and aimed to set up a system similar to the bighorn sheep study on Ram Mountain (Festa-Bianchet et al. 1996, 1998). The need to drug adult goats complicated the capture procedures, however, and when we detected negative effects of capturing and handling goats (Côté et al. 1998a) we curtailed the number of captures. By 1998, the only goats we captured were yearlings, nonlactating adult nannies, and adult males, and even for those sex–age classes we limited captures to a minimum. The situation was partially remedied after we began using remote-display platform scales in 2001, at least in terms of our ability to measure body mass. In the first three years of use, we weighed mountain goats 417 times with the platform scales.

Summary

- Fieldwork on Caw Ridge typically began in mid-May and ended in mid-September each year, but it continued through winter during the first two years of the research.
- We caught mountain goats with traps, tranquilized most adults before handling them, and marked males with ear tags and females with visual collars. A few goats each year were also equipped with radio collars.
- Mountain goats were sensitive to capture and handling; therefore, we curtailed our trapping efforts as the study progressed. Starting in 2001, we weighed marked goats by luring them onto remote-display platform scales.
- Observations of behavior, reproduction, and survival of marked, known-age individuals formed the basis of most of our research.

CHAPTER 4

Home Ranges, Forage Availability, and Habitat Use

Mountain goats on Caw Ridge used a large area of continuous alpine tundra, interspersed with steep drainages with extensive willow bushes and subalpine forest as well as a few cliffs and other steep, broken terrain that could be used to avoid predators, referred to as "escape terrain." Compared to many other areas used by mountain goats, Caw Ridge offered limited opportunities for altitudinal migration because the available altitudinal range of goat habitat was only a few hundred meters. Altitudinal migration allows many other populations of mountain ungulates to follow the "green-up wave" of vegetation phenology, as highly nutritious new vegetation growth becomes available at increasing altitudes as the season progresses.

This chapter provides an overview of how mountain goats used different habitats and different parts of Caw Ridge during summer. Adult males and females had very different home range sizes, possibly reflecting different antipredation strategies. While females moved frequently and roamed over much of the ridge, males mostly remained in one small area all summer. Goat movements and grouping behaviors had important consequences for their social organization, as will be explored in chapter 5. We also present seasonal changes in availability and quality of forage to illustrate two key characteristics of the study area: a very strong seasonality and a rather variable timing of vegetative growth in the spring because of yearly differences in the duration of snow cover. These environmental characteristics have major implications for the timing of parturitions and for population dynamics, which are examined in chapters 7 through 9.

Home Ranges of Males and Females

Similarly to most other sexually dimorphic ungulates (Ruckstuhl and Neuhaus 2002), mountain goats formed two types of groups. Bachelor groups included only adult males, mostly aged three years and older. Nursery groups included females of all ages, kids, and males aged one to four years. The transition of young males from nursery groups to bachelor groups is examined in chapter 5.

Unlike many cervids, whose basic social unit is a group of related females (Albon et al. 1992; Nelson and Mech 1987), mountain goats were not organized as extended families and did not show any obvious subpopulation structuring in area use. All females, kids, and young males ranged over the entire study area, whereas all adult males spent most of the summer within a small area at the western end of the ridge. Consequently, within a given sex–age class, the home ranges of all individuals overlapped. To illustrate this point, fig. 4.1 shows the locations of four adult females during 1996 and 1997. While some individuals were seen more frequently than others in some parts of the ridge, it is clear that no parts of the study area were used only by some goats. In several cases, we saw over 90% of the entire population, excluding adult males, in a single group (see chapter 5). Therefore, the area-use pattern of mountain goats was very similar to that of bighorn sheep (Festa-Bianchet 1986, 1991) and of domestic and feral sheep (Clutton-Brock et al. 1997a; Lawrence and Wood-Gush 1988), with a home-range group of females sharing a common area. An important practical consequence of this behavior was that we could trap or weigh all members of the population at a single trap site.

The summer distributions of nursery and bachelor groups were strikingly different (fig. 4.2). Nursery herds ranged over most of the available habitat (about 28 km^2) for the entire summer, with no obvious monthly differences. Adult males, on the other hand, mostly remained in a small area at the western end of Caw Ridge, particularly in late summer (fig. 4.3). That area-use pattern was repeated every year. The total summer (June–September) home range of nursery groups averaged 21.2 km^2, as determined by the 95% adaptive kernel calculated using the Animal Movement Analyst extension of ArcView GIS (ESRI Inc., Redlands, CA) software and the default LSCV (ad hoc) option. That technique excludes the outermost 5% of locations from the calculation of home range size. From 1995 to 1998, the summer home ranges of the nursery herd varied from 18.1 to 24.8 km^2. During the same period adult males used only 3.6 km^2 (yearly range 2.5–4.6 km^2), an area six times smaller than that used by females. The 80% core areas were much smaller for both sexes,

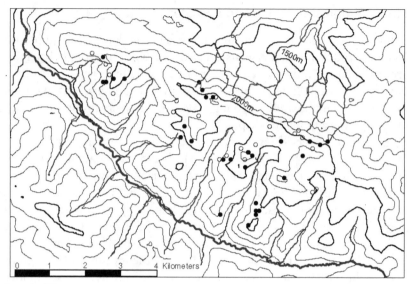

female 27

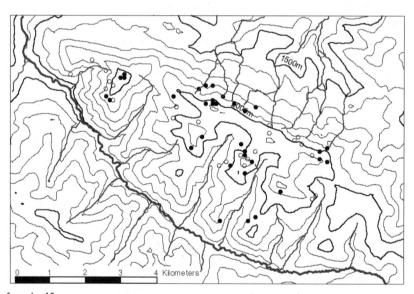

female 46

Figure 4.1. Locations within the Caw Ridge study area of the four adult nannies with the greatest number of sightings from late May to early September in 1996 and 1997. Open circles indicate sightings when all four goats were in the same group, numbers indicate the ID of each goat. These goats were located on average sixty-two times each over the two years, but fewer points are shown because of repeated sightings with the same grid coordinates. Base data provided by Spatial data Warehouse, through Alberta Sustainable Resource Development.

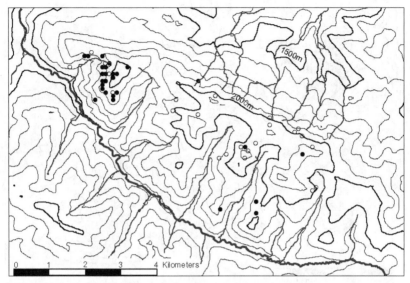

female 64

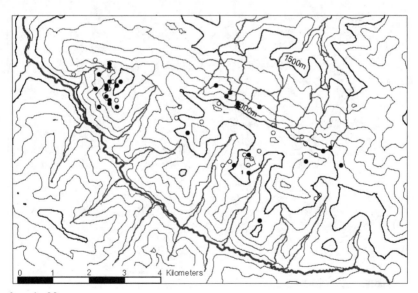

female 93

Figure 4.1. Continued

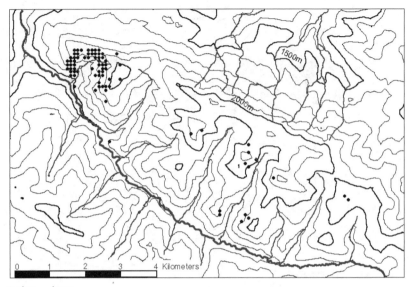

males – June

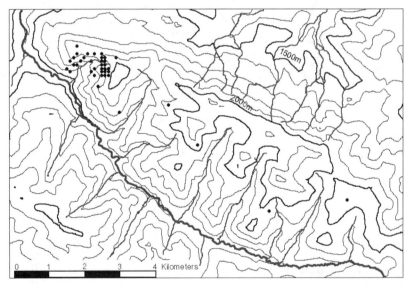

males – July

Figure 4.2. Distribution of mountain goat nursery herds (including females of all ages and males up to three and occasionally four years old) and solitary adult (three years and older) males or bachelor male groups (including only adult males three years of age and older) on Caw Ridge, June to September 1995 to 1998. Base data provided by Spatial data Warehouse, through Alberta Sustainable Resource Development.

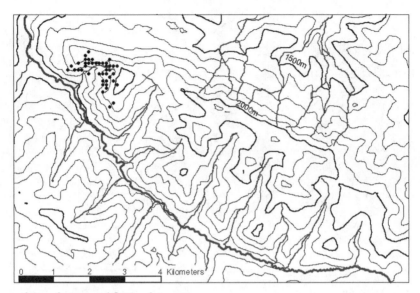

males — August and September

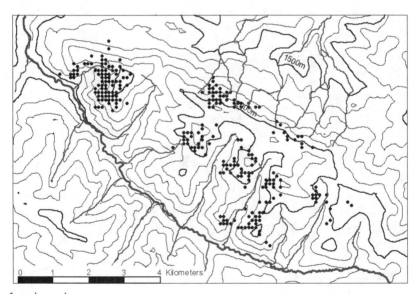

females – June

Figure 4.2. Continued

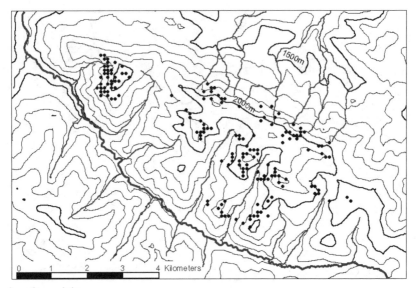

females – July

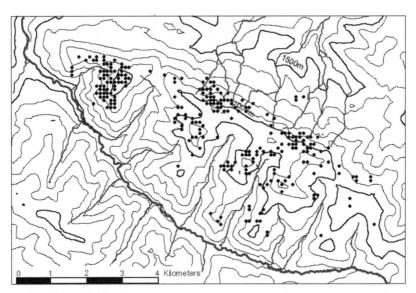

females — August and September

Figure 4.2. Continued

Figure 4.3. "Billy country" at the west end of Caw Ridge, where adult males spent most of their time. Photo by S. Côté.

averaging 8.6 km² for nursery groups (range 3.8–11.1 km²) and 2.1 km² for male groups (range 1.4–2.8 km²). Males were clearly much more sedentary than females during summer. Therefore, mountain goats appear different from most other ungulates, where summer home ranges are either similar for the two sexes, or are larger for males (Mysterud et al. 2001). Chamois, however, show a pattern similar to that of mountain goats: summer home ranges were six to eight times larger for females than for males, with males limiting themselves to areas smaller than 1 km² (Boschi and Nievergelt 2003). In that species, however, some mature males defend territories in summer (von Hardenberg et al. 2000), and therefore their social organization is very different from that of mountain goats.

Dispersion of Parturient Nannies

The strong gregariousness of mountain goat females disappeared briefly around parturition, when females appeared to disperse as widely as possible. In fifteen cases, we saw females that were either giving birth or had just given birth: the kid was wet and very unsteady, or a placenta was visible. Births could occur at any time of the day or night, and some females moved several kilometers in the hours just before parturition. Delivery was quick and was difficult to anticipate when observing females from a distance. Kids suckled and stood within about forty-five minutes. Most mothers cleaned the kid carefully by licking it dry and ate the placenta. Parturient females were either alone or accompanied by their previous year's offspring and remained within a few meters of the parturition site for two to three days. Although it is often assumed that mountain goats select inaccessible cliffs to give birth, only 30% of 141 parturition sites were on cliffs, possibly because of the limited availability of precipitous terrain. As many as 71% of birth sites were within 100 m of treeline and 8% were in the forest.

Several parturient females selected areas where goats were seldom or never observed during the rest of the summer. Other studies of ungulates also reported that parturient females scatter, isolate themselves, and are sometimes found in areas seldom or never used at other times, for example in caribou (Bergerud et al. 1990; Ferguson et al. 1988) and bighorn sheep (Festa-Bianchet 1988d). Wide prepartum dispersion is likely an antipredator adaptation. By distancing themselves from each other, parturient females make it less profitable for a predator to search for them at a time when they are particularly vulnerable. Isolated females may also be more difficult to detect than those in groups. In addition, a period of isolation at birth may be required to establish the mother–kid bond. Parturient females may select sites that are seldom or never used by other mountain goats as a strategy for avoiding both predators and conspecifics during the critical few days after giving birth.

Seasonal and Yearly Changes in Forage Quality

Northern ungulates, including mountain goats, rely on forage with strong seasonal changes in nutritional quality, and consequently show a strong seasonal cycle of gain and loss of body mass (chapter 6). Growing vegetation in spring and early summer has a very high energy and protein content (White 1983). Accordingly, most northern ungulates gain mass during summer. Females also rely heavily on summer forage to support both current and subsequent reproduction: they must produce milk while accumulating enough fat to allow conception in the autumn and sustain gestation over the following winter (Festa-Bianchet et al. 1998). During the short summer, males accumulate both fat reserves and muscle mass, much of which they lose during the rut (Pelletier 2005; Yoccoz et al. 2002). In winter, however, the nutritional quality of forage is very low. For five to eight months each year, depending on latitude and altitude, ungulates are in a negative energy balance and lose mass (Festa-Bianchet et al. 1996).

Generally, the period when vegetation quality is insufficient for maintenance lengthens with increasing latitude or elevation. At Caw Ridge, the growing season begins very late. In some years, snow cover until late June prevents vegetative growth over most areas used by goats. All vegetation growth appears to stop by mid-August, when the ridge turns yellow to brown. Winter conditions sometimes set in by September: temperatures drop well below zero and snow cover may last for several days. Consequently, mountain goats on Caw Ridge had only three to four months each year to accumulate the metabolic resources required for survival, growth, and reproduction, with important consequences for the physical development of growing individuals and for population dynamics.

Because the timing of the rut is affected by photoperiod, it probably does not vary much from year to year. Ungulates have little flexibility in the duration of gestation (Berger 1992; Byers and Hogg 1995) and the timing of births in mountain goats varies little from one year to another (chapter 7). Differences in the initiation of vegetative growth in spring should therefore affect the growth and survival of neonates because mothers may be unable to produce sufficient milk when feeding on low-quality overwintered forage. Late snowmelt can also increase mortality, particularly of juvenile and senescent ungulates, because it prolongs the period during which animals must rely partly on stored fat (Jacobson et al. 2004; Taillon et al. 2006). If snowmelt is late, fat resources may be depleted before new forage becomes available, and animals may die of starvation (Dumont et al. 2000).

The timing and duration of access to highly nutritious forage are also affected by topography. Vegetation growth starts earlier at low elevations and moves to higher elevations as the summer progresses, following the same pattern as snowmelt. Ungulates that migrate seasonally over an altitudinal gradient can therefore prolong access to highly nutritious growing forage (Festa-Bianchet 1988d; Langvatn et al. 1996). The altitudinal range exploited by the Caw Ridge mountain goats was only approximately 400 m. On slopes of similar aspect, forage growth would start (and peak) only about six days later at 2100 than at 1700 m (Hoefs and Cowan 1975). Elevation, however, is not the only landscape variable affecting the timing of vegetative growth in the mountains. Other important factors include aspect and snow accumulation. Snow melted later on north-facing than on south-facing slopes, and some areas had deep snow accumulation because of windblown deposition during winter. There was new vegetative growth somewhere on Caw Ridge until early August, but after that date melting snowbanks left bare soil in their wake.

BOX 4.1
Using Fecal Crude Protein to Measure Forage Quality

Factors that affect fecal crude protein values in ruminants include the ash content of feces (Wehausen 1995) and the amount of secondary compounds, particularly tannins, in the forage (Robbins et al. 1987). Fecal ash is primarily made up of grit and dust ingested while feeding. The occasional low crude protein content of some fecal samples collected in summer could be due to grit ingested while licking mineral-rich soil that increases the specific weight of fecal pellets. Indeed, "ash-free crude protein" better reflects the nutritional value of forage (Wehausen 1995). Unfortunately, most fecal samples collected at Caw Ridge were not analyzed for ash content. For forty-five fecal samples collected between 1990 and 1992, ash content varied from 12.6 to 36.6% (average ± SD = 18.2 ± 4.76) and appeared negatively correlated with fecal crude protein ($r = -0.28$, $P = 0.07$) as expected if large amounts of grit and dust in fecal pellets increased specific weight. Although variability in ash content contributed to the measurement error of crude protein, it should not have produced any systematic bias and should not have varied among years.

If a ruminant eats plants with a high content of phenolic compounds, fecal crude protein may increase without a corresponding increase in protein available for digestion, since phenolics bind with proteins and make them indigestible (Robbins et al. 1987). Compared to grass, browse generally has a higher content of phenolics and other secondary compounds. Ruminants that are primarily grazers, such as bighorn sheep, eat little phenolic-rich forage. The protein content in fecal samples of bighorn sheep provides a good measure of the quality of the forage recently consumed (Blanchard et al. 2003). Mountain goats are considered mixed-feeders and consume both coniferous and deciduous browse (Adams and Bailey 1983; Dailey et al. 1984; Fox and Smith 1988). Summer diet in any specific site varies with forage availability but is dominated by grasses (Laundré 1994; Rideout 1974). Laundré (1994) summarized ten studies of feeding habits of mountain goats and found that summer diet included on average 52% grass, 30% forbs, and only 16% browse. Consequently, the effects of secondary compounds on fecal crude protein should be limited. Yearly differences in fecal crude protein are likely to reflect yearly differences in forage quality, rather than differences in either ash content or consumption of phenolic-rich forages (Dailey et al. 1984).

To monitor forage quality, we analyzed goat fecal samples for crude protein content. Although subject to a number of biases (Hobbs 1987; Robbins et al. 1987), the crude protein content of ungulate feces provides a useful index of seasonal and yearly changes in nutritional quality of forage (Kamler and Homolka 2005; Leslie and Starkey 1987) because it reflects what animals have eaten. We relied on fecal crude protein to detect seasonal and yearly changes in the nutritional quality of forage eaten by mountain goats.

Seasonal and yearly changes in fecal crude protein (fig. 4.4) revealed several important aspects of the nutritional ecology of mountain goats. First, although in general the best forage was available from mid-June to early August, the seasonal pattern of protein content varied substantially from year to year, as did the duration of the period when fecal crude protein remained near its peak. Second, the initiation of forage growth in most years was rather late. It was typically two to four weeks later than in our bighorn sheep study areas located a few hundred kilometers to the south (fig. 4.5). The protein content of fecal samples typically peaked in

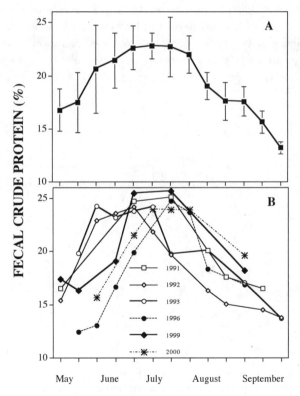

Figure 4.4. (a) Mean and standard deviation of crude protein content of mountain goat fecal samples collected every ten days from late May to late September, 1991 to 2000. (b) Interannual variation in seasonal values of fecal crude protein. The figure shows six of ten years of available data.

mid- or late July. In some years, such as 1996 and 2000 (fig. 4.4) it appeared that little vegetation growth had taken place by mid-June, when goats were still consuming mostly low-quality, overwintered forage. In other years, such as 1993, forage quality peaked in early June and declined by late July.

A comparison of seasonal fecal crude protein levels at Caw Ridge with those of two bighorn sheep study areas in Alberta (fig. 4.5) revealed how late spring arrived for the Caw Ridge mountain goats. Late springs mean long winters and an extended time when animals must rely on stored reserves because the forage cannot satisfy daily dietary requirements. In late May and in June, reproductive females also need to satisfy the high energetic demands of lactation. On average, the peak in vegetation quality at Caw Ridge was one month later than at Ram Mountain, at similar elevation about 250 km to the south (Blanchard et al. 2003), and two to three weeks later than in alpine areas at Sheep River, about 450 km to the south but at elevations 300 to 400 m higher than Caw Ridge. Bighorn sheep at Sheep River also have access to a low-elevation winter range, where they can feed on new vegetation growth about six weeks earlier than the Caw Ridge goats (Festa-Bianchet 1988d). On the other hand, in

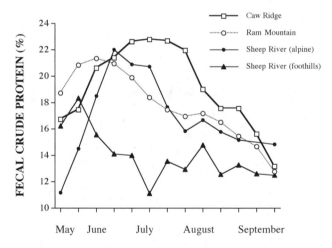

Figure 4.5. Mean crude protein content of fecal samples from mountain goats at Caw Ridge (1991–2000) compared with samples from bighorn sheep at Ram Mountain and at Sheep River, by ten-day periods from late May to late September. At Sheep River, most ewes move from the foothills range to the alpine range in late May, and return to the foothills range in August and September.

most years fecal crude protein from mid-July to August was substantially higher at Caw Ridge than at either Sheep River or Ram Mountain (fig. 4.5), possibly because of greater summer precipitation. Peak fecal crude protein also appeared higher at Caw Ridge than in the two bighorn sheep study areas, but that difference may be due to interspecific differences in digestion or in phenolic content of the diet, since mountain goats generally utilize more browse than bighorn sheep (Laundré 1994).

Given the many variables affecting the relationship between fecal crude protein values and forage quality, fig. 4.5 must be interpreted cautiously. Small differences between areas may be due to differences in phenolic or ash content of local diets or to differences in digestive metabolism rather than to differences in forage quality. Two patterns, however, are clearly evident: compared to bighorn sheep at both Sheep River and Ram Mountain, mountain goats at Caw Ridge experience a later initiation of vegetative growth but enjoy longer access to high-quality forage. These two characteristics of plant phenology are key to many aspects of the growth and survival of young mountain goats examined in later chapters.

Forage crude protein content was measured by Martine Haviernick in 1993 and 1994 at several sites on Caw Ridge. The seasonal pattern of crude protein in graminoids and willow was similar to that in fecal samples (fig. 4.6), peaking in early summer and then decreasing. Forage protein content also varied substantially from one part of Caw Ridge to another (Haviernick 1996).[1] There were, however, no obvious explanations for local differences in forage protein. Because site NS was on the north side of Caw Ridge, it was expected to show a later protein peak than the other sites, but that was not obvious (fig. 4.6). Site WES was located in the area most often used by adult males during the summer. It appeared to have the lowest graminoid crude protein content in 1993, but in 1994 its level of crude protein was comparable to that of other sites.

Given that different parts of Caw Ridge reached their peak forage quality and quantity at different times, we expected that over the summer mountain goats would select the sections of the ridge that offered the best nutrition. There was no evidence, however, that seasonal use of different parts of Caw Ridge was determined by differences in forage protein. During two years of research, Martine Haviernick determined that goats did not select sites according to seasonal differences in peak forage protein or in forage biomass (Haviernick 1996). Consequently, the dispersion of nursery herds could not be explained by local differences in forage quality or quantity.

Forage Consumption

In most years from 1994 to 2002, we clipped 20 × 20 cm vegetation quadrats inside and outside a series of 1 × 1 m exclosures on a site at the west end of Caw Ridge that was regularly used by mountain goats in both summer and winter. Quadrats were clipped in the first week of June and in the first week of September. We collected all vegetation over 1 cm high within the quadrats, but often there was little or no dead vegetation in the samples. Therefore, the June sample reflected early spring growth, and the September sample measured the summer biomass accumulation that had not been removed by mountain goats or other herbivores. Rodents, lagomorphs, or insects could consume vegetation inside the exclosures. Vegetation samples were oven dried and weighed. There was al-

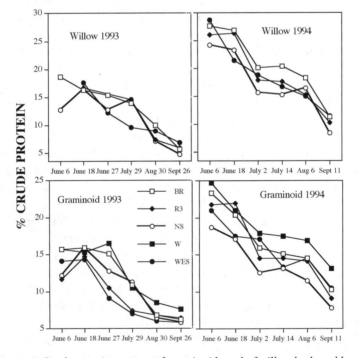

Figure 4.6. Crude protein content of graminoids and of willow buds and leaves in five different sites regularly used by mountain goats on Caw Ridge in 1993 and 1994. There was no willow on site W, and willow was not sampled at all sites in 1993 (from Haviernick 1996).

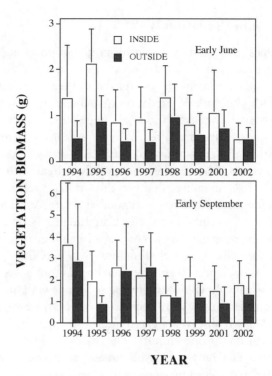

Figure 4.7. Dry vegetation biomass ($\bar{X} \pm$ SD) collected from 20 × 20 cm quadrats inside and outside 1 m^2 exclosures on the west end of Caw Ridge in early June and early September, 1994 to 2002. Note change in scale on *y*-axis.

most twice as much forage biomass in September as in June (fig. 4.7). In all years, there was more biomass inside than outside the exclosures in June (average difference was 40%), suggesting that goats started feeding on new vegetation as soon as it became available. With the exception of 1997, there was also more biomass inside than outside the exclosures in September (average difference of 22%), suggesting an impact of mountain goats on vegetation biomass. Overall our limited data on summer forage availability do not suggest that goats exerted a very high grazing pressure. The site we sampled was heavily used by the goats, yet vegetation biomass accumulated during summer. There were no significant correlations between the number of adult females or the total number of goats and either vegetation biomass outside the exclosures or the difference in biomass inside and outside the exclosures.

Habitat Use and Predation Risk

The many types of habitat available to mountain goats on Caw Ridge varied in elevation, slope, aspect, forest cover, and soil drainage. We compared the use of open and forested habitats because forested habitats may involve a higher risk of predation. Mountain goats rely upon the typical antipredator strategy of mountain ungulates, being vigilant to visually detect approaching predators. When a predator is detected and an attack is possible, the goats move to very steep or rocky escape terrain where predators would be unable to either follow them or to attack without substantial risk of injury. Because this strategy relies on sighting predators from a distance, it requires good visibility. At least two of the major predators of mountain goats on Caw Ridge rely on cover to approach their prey: cougars and grizzly bears. Nine predation attempts by grizzly bears were observed, including four in 1996 (Côté and Beaudoin 1997) and five more since. Bears attempted to capture goats by running toward them, usually downhill, and trying to hit one with their forepaws. None of the attacks we saw were successful, but on the morning of September 12, 1996, we observed a grizzly that had apparently just killed a young adult male on the open tundra. The male was alive and healthy the evening before. The only successful predation attempt we witnessed (Côté et al. 1997a) involved two wolves that ran down a yearling after approaching to within about 200 m behind a ridgeline. On another occasion, a lone wolf caught a kid by a leg, but released it after being horned by the kid's mother (Côté et al. 1997a).

Almost all documented cases of successful predation were in forested habitat or within 50 m of forest cover (table 4.1). Therefore, use of forested areas appeared to involve high predation risk. As expected, goats tended to be more alert in forested than in open habitats, although the difference was not very consistent. In 1993, active females spent approximately 10% of their time alert in forested habitat, and 5% in open habitat. In 1994, the trend was the same but the difference was not significant: females spent approximately 6% of their active time alert while in the forest and 4% in open areas (Haviernick 1996).

Given an apparently higher risk of predation, one would expect goats to avoid forested areas, unless there was some essential resource that was only available in the forest. Habitat use by goats on Caw Ridge differed substantially according to group type: nursery groups were seen in forested habitat only 8.8% of the time but male groups were seen in forested areas 45% of the time (fig. 4.8). Clearly, it is much easier to see goats in open than in closed habitats, but the sightability bias should be stronger

TABLE 4.1
Known Predation Events on Mountain Goats on Caw Ridge

Date	Sex–age of goat	Predator	Habitat type
September 89	Male kid	Wolf	Open forest
October 89	Female kid	Grizzly bear	Open forest
November 89	Male kid	Wolf	Tundra
February 90	Male kid	Unknown	Closed forest
September 90	Male kid	Wolf	Open forest
October 90	Female kid	Grizzly bear	Open forest
October 90	Male kid	Grizzly bear	Krummholtz
January 91	Female kid	Cougar	Open forest
June 91	Male yearling	Grizzly bear	Krummholtz
July 91	Female yearling	Cougar	Krummholtz
September 91	Female kid	Cougar	Open forest
September 91	Female yearling	Grizzly bear	Open forest
February 92	Female kid	Wolf	Open forest
September 92	Female kid	Grizzly bear	Tundra
August 95	Female yearling	Wolf	Tundra
September 96	Adult male	Grizzly bear	Tundra
September 96	Male kid	Wolf	Tundra
May 99	Adult female	Wolf	Tundra

Includes habitat where the predation took place. The high number of kids with known causes of death in the first three years of the study corresponds to the period when we used mortality-sensitive radio collars and attempted to monitor each collar daily. It does not reflect higher predation in those years.

for males that were often in small groups than for nursery groups that were much larger (chapter 5). It is therefore likely that the actual use of forested areas by adult males was greater than our estimate. Neither sex showed a strong seasonal pattern in habitat use over the summer (fig. 4.8).

If mountain goats are at greater risk of predation in closed than in open habitats, why did males spend about half their time in forested areas? Mature males spent most of the summer within a very restricted area (fig. 4.2) and were typically seen either in small grassy openings within the forest or near treeline. They often foraged in the forest and rested just above treeline. The area used by males at the western end of Caw Ridge was at lower elevation than most of the areas used by nursery groups during summer. Although the difference in elevation was only a few hundred meters, the protection from wind and increased moisture availability offered by the forest cover resulted in abundant green forage

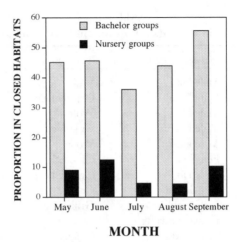

MONTH

Figure 4.8. Use of forested habitats by mountain goat bachelor and nursery groups on Caw Ridge from 1995 to 1999. Based on 1537 sightings of nursery groups (monthly range 144 to 593 sightings) and 361 sightings of bachelor groups (monthly range 31 to 147).

from mid-June to August. In addition, this was the only section of Caw Ridge where aspen was readily available for browsing. During summer, mature males appeared to do little else than eat, ruminate, and sleep. Perhaps they had no reason to move out of the section of Caw Ridge that appeared to have the most forage. On the other hand, their behavior could have made them extremely vulnerable to stalking predators: males spent most of the summer within a small area, much of it under forest cover. Adult male distribution was predictable for us, and presumably it was also predictable for potential predators.

Sex Differences in Daily Movements

Nursery groups made extensive daily movements during summer, sometimes covering 5 to 9 km in a day. In contrast, adult males tended to remain within a restricted area (fig. 4.2). In most years, males made one or two excursions to the trap site, often at night, and were occasionally seen elsewhere on the ridge (fig. 4.2).

To estimate how far goats traveled during a day, we examined the sightings of the ten adult females and up to ten adult males that were seen most frequently each year from 1995 to 2003. We calculated distances between UTM coordinates when the same goat was seen in consecutive days. On average, the linear distance between sightings of the same adult female in successive days from June to September was 116 ± 112 m, compared to only 50 ± 34 m for adult males.

Why Are Nursery Herds So Mobile?

Nursery herds could be found almost anywhere on the ridge in summer (fig. 4.2). There were no clear, repeatable seasonal patterns in area use. For example, the proportion of nursery groups seen at the west end during different months was not consistent across years (fig. 4.9). Summer movements of goats were apparently not due to local seasonal differences in the availability or quality of forage. Goats presumably used areas where they could find good forage, but it seems unlikely that they needed the entire 20 to 25 km^2 that they used in most years simply to satisfy their food requirements.

While nursery groups moved about the ridge in a somewhat unpredictable fashion, adult males remained within a small area and were much more predictable in their spatial distribution. This sexual difference suggests that mountain goat movements during summer could not be explained by a need for food. Because males are larger than females, males should require more forage. Therefore, if home range size were dictated by food requirements males should use larger areas than females, the opposite of what we found.

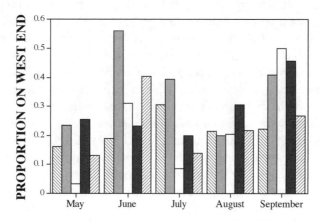

Figure 4.9. Proportion of nursery groups seen in the west end area of Caw Ridge from 1995 to 1999. Each pattern of shading represents one of five consecutive years.

A similar area-use behavior during summer has been described for bighorn sheep: ewes move greater distances than adult rams while foraging and are less likely to remain in the same area on consecutive days (Ruckstuhl 1998). We suggest that the frequent movements and somewhat unpredictable distribution of female mountain goats are an antipredator strategy, not a foraging strategy. By moving in an unpredictable fashion over a wide area, goats may make themselves less valuable for predators, which would have to spend more time searching than if goats consistently used a small area. Female groups are likely more sensitive than male groups to predation risk because kids and yearlings are more vulnerable than other sex–age classes. Studies of other ungulates also suggest that tolerance to predation risk is lower for females than for males (Bleich et al. 1997).

The main predators of mountain goats on Caw Ridge are wolves, grizzly bears, and cougars. All three species can kill ungulates much larger than mountain goats (Berger et al. 2001; Larsen et al. 1989; Ross and Jalkotzy 1996). Although goats may sometimes successfully defend themselves against predators, there is no evidence that males could be more successful in doing so than females, or that an attack on an adult male presents a greater risk to a predator than an attack on an adult female. Consequently, if nursery groups travel to avoid predation, then males in the much more sedentary bachelor groups should experience greater predation rate. Unfortunately, we have insufficient data on causes of death to test this prediction. The higher age-specific mortality rate of adult males than of adult females (chapter 9) may be due to a greater predation risk, but it may also ultimately be due to differences in reproductive strategy rather than habitat use. On the other hand, even a long-term study may not necessarily be able to statistically confirm an association between behavior patterns and predation risk. Predation is extremely difficult to monitor while studying prey species. We could rarely determine causes of death. In addition, the sample of mature males that died during our fifteen-year study was only twenty-three, because many males died or emigrated before age four (chapter 9). At least ten (43%) adult males disappeared between late May and the end of August, but their cause of death was unknown. The proportion of females aged four years and older that disappeared during the same period was 33% ($n = 51$). Although the difference is not significant, the trend for higher summer mortality in males than in females could be related to the use of riskier habitats by males. We do not know if sexual differences in habitat selection persist during winter.

Conservation Implications of Goat Habitat-Use Patterns

Female goats appear to roam over a large area as an antipredator strategy. If some of their habitat were destroyed or rendered unusable by human activities, the goats would be forced into a smaller area. They would then be more spatially predictable and therefore would likely be at a greater risk of predation. A reduction in available habitat could increase mortality even if the amount of food available in the smaller area were sufficient to support the population. In other words, the carrying capacity for mountain goats may not simply depend on the amount of available forage but may also be affected by the availability of sufficient suitable habitat for the goats to maintain an unpredictable dispersion. The limited escape terrain on Caw Ridge compared to other areas used by mountain goats may increase the selective advantage of frequent daily movements. If goats predictably foraged in areas near the best escape terrain, they could be vulnerable to ambush predators such as cougars. We expect that goats in areas with more extensive escape terrain will have smaller home ranges and make less frequent movements because they could spend much of their time in areas inaccessible to predators. If we are correct in our interpretation that nursery herds are highly mobile as an antipredator strategy, then both the carrying capacity of mountain goat habitat and home range sizes in different populations could be somewhat independent of summer forage availability and therefore extremely difficult to predict. Within a certain range, yearly changes in forage availability may have little or no effect on mountain goat population dynamics. Mountain goats are highly susceptible to industrial activities (Joslin 1986), possibly in part because they suffer increased predation when prevented from using certain sections of their traditional range.

Summary

- During summer, nursery groups of females, kids, and yearlings ranged over a much wider area than bachelor groups of adult males. All females used all parts of Caw Ridge and there was no evidence of spatial structuring within the population.
- Parturient females isolated themselves from other goats and dispersed widely over the ridge, possibly to make themselves more difficult for predators to find.
- Fecal crude protein suggested that the timing of initiation of vegetation growth on Caw Ridge was later and more variable than in

other mountainous areas in Alberta, likely due to variability in timing of snowmelt. The period during which mountain goats could feed on high-quality vegetation, however, appeared longer than in other study sites.

- The frequent movements and large summer home range of females were unlikely to be explained by their food requirements or by selection of sites with the best forage. They may have been part of an antipredator strategy to make the location of nursery groups unpredictable.

Statistical Note

1. Accounting for sampling date, differences in protein content among sampling sites were significant in both years and for both types of forage (nested ANOVAs, $P = 0.012$ for willow in 1993; $P < 0.001$ in all other cases).

CHAPTER 5

Social Organization

The social structure of a population can affect patterns of resource use, vulnerability to predation, and the transmission of genetic variability from one generation to another (Clutton-Brock et al. 1997a). It is therefore important to understand what factors affect the social organization of a population. Social organization, however, is not an individual attribute, but depends on how individuals respond to the behavior of other individuals in the population. Social organization should therefore be flexible and should vary among and within populations following changes in density, kin relationships, and sex–age structure. Because it is not an individual characteristic, social organization itself cannot be subject to natural selection: selective pressures can instead affect how an individual may behave in a given social context so as to maximize its fitness. In this chapter we will describe the social organization of mountain goats and explore its implications for population dynamics and for the conservation of this species.

Mountain goats at Caw Ridge were highly gregarious. Their social system included sexual segregation and frequent postweaning mother–offspring associations but no preferential association among related adult females. Here we first describe how group size varied according to group type and season, then explore the development of sexual segregation, mainly by examining the behavior of young males as they switch from nursery to bachelor groups. We then use data on known kin relationships to establish whether or not mountain goat females associate with relatives.

Dominance relationships among mountain goats at Caw Ridge have already been examined (Côté 2000; Côté and Festa-Bianchet 2001c), and here we will only provide a short synthesis of previous publications while concentrating on group composition and structure. Adult mountain goat females are frequently aggressive to each other and maintain a stable and linear social hierarchy, based on dyadic relationships that vary little from year to year (Côté 2000; Fournier and Festa-Bianchet 1995). When females interact, they usually avoid contact, possibly because their sharp horns can easily inflict injuries if interactions escalate. It is likely advantageous for each individual to know her place in the social hierarchy and to avoid challenging dominant goats, given the potentially high cost of a stab wound (Geist 1967). Instead, subordinate females typically retreat immediately upon being challenged by a dominant individual, and almost all interactions are initiated by the dominant member of a pair. Social rank increases very strongly with female age: when two adult females of different ages interact, the older one wins 94% of the time (Côté 2000) so that the oldest females are typically the most socially dominant.

Size and Composition of Mountain Goat Groups

Like many other ungulates, mountain goats are sexually segregated during the summer, forming two types of groups. Bachelor groups were composed only of males aged three years and older, while nursery groups included females of all ages and males up to the age of four years. Occasionally, older males were seen in nursery groups and approximately 1% of sightings of two-year-old males were in bachelor groups. As presented in chapter 4, during the summer, bachelor and nursery groups differed in habitat use, home range size, and daily movement rates. Although nursery and bachelor groups very seldom mixed, most areas used by bachelor groups at the west end of Caw Ridge were also used by nursery groups. The only exception was that some of the more heavily forested areas used by groups of males were rarely utilized by nursery groups. Sexual segregation was therefore temporal but not necessarily spatial, similarly to bighorn sheep (Ruckstuhl 1998).

Inevitably, because the population included many more females, yearlings, and young males than adult males, bachelor groups were much smaller than nursery groups. Between 1995 and 1999, mean bachelor group size was 2.6 ± 2.2 individuals ($n = 374$ groups), while nursery groups averaged 11.9 ± 18.1 goats excluding kids ($n = 1602$ groups). Nursery groups were small in late May, when parturient females isolated themselves to give birth, increased through the summer, then declined slightly in September (fig. 5.1). In July and August, we often saw over

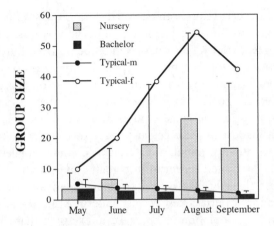

Figure 5.1. Monthly variation in size of nursery and bachelor groups of mountain goats on Caw Ridge, 1995 to 1999. Histograms report mean group size with SD; lines indicate the typical group size (mean group size experienced by individual goats, see Jarman 1974) for males in bachelor groups (thin line, closed circles) and for females in nursery groups (thick line, open circles). Based on 359 bachelor and 1493 nursery groups (including solitary individuals but excluding kids of the year).

90% of the population, excluding adult males, in a single group, sometimes including over a hundred goats. Bachelor groups followed an opposite seasonal trend, gradually decreasing in size from May to September (fig. 5.1). With the exception of parturient females in late May, mountain goats were very seldom alone. Adult males were three times as likely to be solitary than adult females: 9.6% of adult male sightings and 3.3% of adult female sightings were of single individuals.

Mean group size, however, is an observer-centered and somewhat biologically misleading measure of gregariousness, particularly when group size varies substantially. A lone goat and a group of fifty-nine would produce a mean group size of thirty. Because all but one goat experienced a group size of fifty-nine, however, the mean group size would be meaningless. Mean group size is heavily influenced by small groups and underestimates the group size experienced by most animals, which is better represented by "typical group size" (Jarman 1974), the mean group size experienced by each individual. In the example above, fifty-nine goats experienced a group size of fifty-nine and a single goat a group size of one, for a typical group size of fifty-eight. For bachelor groups, the observer-centered mean group size and the animal-centered typical size were not very different and both declined during the summer (fig. 5.1). For nursery groups, however, typical group size was much larger than mean group size. From July through September, most adult females experienced groups of forty or more. Typical group size provides a much better documentation than mean group size of the strong gregariousness of female mountain goats at Caw Ridge during summer.

In August, most of the nursery population was often seen in a single group, which made complete censuses easy to do. Between 1995 and 1999, the total nursery population (yearlings, adult females, and two-year-old males) averaged seventy-three goats (range sixty-six to eighty-four) in August, therefore the typical group included approximately 75% of the entire nursery herd. These results underline the lack of spatial substructuring in the population (chapter 4), and suggest that each nursery goat had frequent opportunities to interact with all other nursery goats in the population. In contrast, the typical group of bachelor males peaked at 5.3 goats in May and declined to 1.9 by September, suggesting that males became less gregarious as the rut approached. Given that the population included on average fourteen males aged three years and older (range nine to sixteen) between 1995 and 1999, as the summer progressed the average adult male associated with a decreasing proportion of other males in the population, from 33% in May to 7% in September. Clearly, female and juvenile mountain goats on Caw Ridge were much more gregarious than adult males during summer. Sexual differences in gregariousness could be explained by differences in predation risk and possibly in habitat selection. Nursery herds were typically seen in open areas, often several hundred meters away from escape terrain, where large groups may have increased the ability to detect predators. In contrast, bachelor groups spent about half of their time in forested areas (chapter 4), where large groups may be less advantageous than in open habitats for predator detection. Many ungulates form larger groups in open areas than in closed habitats. This relationship applies across species (Jarman 1974) as well as within species: roe deer, fallow deer, and moose all form larger groups in open habitats than in forested areas (Apollonio et al. 1998; Clutton-Brock et al. 1988; Gerard et al. 1995; Miquelle 1990; San José et al. 1997). Cohesion of group members is easier in open habitats, where animals can see each other (Gerard et al. 1995), than in forests, where visual barriers may lead to frequent group splitting. In areas where mountain goats make greater use of precipitous cliffs, they should form much smaller groups than those we saw in the extensive open tundra of Caw Ridge, due to a combination of smaller areas available for foraging and lower predation risk.

The Development of Sexual Segregation

Adult male and female mountain goats form separate groups outside the rut. Sexual segregation is a widespread phenomenon in ungulates and in many other mammals. Several explanations for sexual segregation in ungulates have been proposed (Main et al. 1996; Ruckstuhl and Neuhaus 2002). Theories with the most empirical support can be grouped into four categories that attribute segregation to sexual differences in (1) predation risk or predation avoidance strategy, based mostly on the assumption that females and juveniles are more vulnerable to predation than males; (2) diet choice and therefore habitat selection, typically assuming that, compared to females, males can survive on lower-quality forage because of larger body size; (3) sociality, with each sex being more interested in forming groups with other members of the same sex because of behavioral compatibility; and (4) time budgets: due to dimorphism in body size, males and females have different optimal schedules of time spent foraging and ruminating and therefore mixed-sex groups cannot be cohesive. An earlier suggestion that males were forced into suboptimal habitat by their inability to compete with females on short grass swards (Clutton-Brock et al. 1987) has been tested and rejected (Conradt et al. 1999). Recent research and reviews of the literature suggest that sexual segregation in ungulates can rarely be explained by a single variable (Bowyer 2005). Generally, the best explanation for why the sexes seldom mix outside the mating season is a combination of theories one and four: males and females differ both in time budgets and in either actual predation risk or the willingness to accept risk. Males may accept a higher risk of predation to obtain more or better food, because the potential fitness pay-offs of large body size are greater for males than for females (Conradt 1998; Ruckstuhl and Neuhaus 2002).

At Caw Ridge, males were often difficult to observe for extended periods because when foraging they entered the forest and disappeared from view. We therefore do not have enough information on time budgets to properly test the idea that sexual segregation is inevitable because the foraging/ruminating cycles of males and females are different (Ruckstuhl 1998). An interspecific comparison of body mass dimorphism with sexual differences in time budgets suggests that female mountain goats should spend about 20% more time foraging than males (Ruckstuhl and Neuhaus 2002). If that were the case, it would be very difficult for mixed-sex groups to remain cohesive, because at some point adult males would lie down to ruminate while adult females would continue foraging and would move away.

BOX 5.1
Measuring Aggressiveness in Mountain Goats

The number of aggressive interactions experienced or even initiated by an animal should depend on the number of opportunities for aggression. An animal in a small group may not be any less aggressive than one in a large group, but it will likely have fewer opportunities to interact. Group size, however, provides a very imperfect metric of opportunities to interact, because these can vary with group geometry. For example, when foraging, mountain goats often distribute themselves on an irregular line that moves slowly forward. A goat in a group of ninety may only have two neighbors with whom it could potentially interact. We measured aggressiveness as the ratio of the number of interactions over the number of opportunities to interact (Côté 2000). During 30-min focal-animal observations (Altmann 1974) we noted all aggressive interactions between the focal animal and other goats. Observations were conducted from June 1 to September 15, 1994–1997. Based on preliminary observations of agonistic interactions, we considered that when another goat was within 4 m of the focal animal, regardless of which goat initiated the approach, there was an opportunity for interaction. We divided the number of interactions initiated by the focal individual by the total number of opportunities. Because older goats almost always win interactions against younger individuals, we only considered the aggressiveness of the focal individual toward other goats of the same age or older. Aggressiveness was therefore independent of group size and could range from 0 when all goats within 4 m were tolerated to 1 when an aggressive interaction occurred every time another goat was within 4 m (Côté 2000). We conducted 96, 21, and 31 observations of adult females, adult males, and two-year-old males, respectively, in which the focal individual was active for more than 20 min and had at least one opportunity to interact

Aggressiveness did not differ between years, and lactating and nonlactating females had similar levels of aggressiveness toward females of the same age and older (Côté 2000a). Adult females were twice as aggressive (0.052 ± 0.015[SE]) as adult males (0.026 ± 0.017), but the difference was not significant ($z = -0.54$, $P = 0.4$). Two-year-old males showed intermediate aggressiveness values (0.041 ± 0.014), but did not differ statistically from adult females ($z = -1.11$, $P = 0.2$) or adult males ($z = -1.28$, $P = 0.1$). Nevertheless, these data suggest that sexual segregation could not be explained by females avoiding aggressive males.

It has also been proposed that sexual segregation in ungulates may develop because females avoid associating with the more aggressive males (Bon et al. 2001). In mountain goats during summer, however, females appear to be the most aggressive sex. (See box 5.1.)

Aggression by adult females does not explain why males leave nursery groups, because female aggressiveness toward young males decreases as males age (Romeo et al. 1997). In addition, Romeo et al. (1997) showed that two- and three-year-old females that remained in nursery groups received more threats than males of the same age. Just before they leave nursery groups, most young males are dominant over adult females: from mid-July to mid-September 1994–1997, three- and four-year-old males won 68% of 124 interactions with adult females. Therefore, young males leave nursery groups by choice, not by force.

The behavior of young males may provide clues to the sexual segregation found among adult mountain goats. The switch in group type occurred over several years (fig. 5.2) and varied between individuals. The proportion of sightings of individual three-year-old males in nursery herds varied from 16 to 100%, while the range for four year olds was 0 to 73%. Only two of sixteen four year olds were never seen in nursery herds, while five of twenty-six three year olds were never seen in bachelor groups. All two-year-old males were seen in nursery herds at least 93% of the time and 78% of them were seen only in nursery groups. Beginning at age four, males were most often seen in bachelor groups (fig. 5.2). Males aged four and five years were seen alone 13% of the time. Males older than six years were solitary in 16% of sightings.

If young males switched from nursery to bachelor groups for social reasons (either because they were chased from nursery groups or because

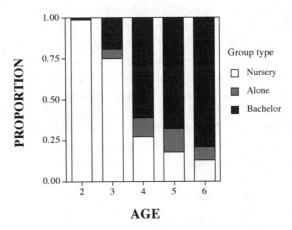

Figure 5.2. Group choice by young males at Caw Ridge, June to September 1995 to 1999. Bars indicate the proportion of sightings in different group types. Only males seen more than ten times during a summer were included. On average, each male was seen thirty-one times a year. Sample size decreased from twenty-eight two year olds to five six year olds, and included thirty-four different individuals.

they preferred to interact with other males), they may simply switch once and then stay in bachelor groups. If, on the other hand, young males moved back and forth between the two types of group, then perhaps they were sampling them and may not have had a strong preference for either group type. We therefore investigated group switching by males aged three and four years that were seen in both types of groups within the same year. All but seven of twenty-five males switched between nursery and bachelor groups at least three times, and males switched group type on average four times over the summer. One three year old switched group type at least ten times, and a four year old switched groups seven times. That behavior suggests that young males were sampling group type and did not immediately prefer one to the other. Young males preferred bachelor groups more (or nursery groups less) as they aged (fig. 5.2). The switch from nursery to bachelor groups was gradual, and most males moved back and forth between group types over a period of two to three years.

Ruckstuhl and Neuhaus (2002) suggested that sexual segregation in ungulates is a consequence of sexual dimorphism. Rumen size increases isometrically with body mass, while metabolic requirements increase more slowly than body mass. Consequently, relative to metabolic requirements, a large ungulate has a larger rumen volume than a small ungulate. Therefore, the schedule of foraging and ruminating should differ among ungulates of different body size, both interspecifically and intraspecifically. In bighorn sheep, adult males spent less time grazing, had shorter grazing bouts, and spent much more time lying down than adult females (Ruckstuhl 1998). Bighorn rams are about 40 to 50% heavier than ewes (Festa-Bianchet et al. 1996). Ruckstuhl argued that males have different time budgets than females, and that individuals could not synchronize their activities to those of other group members unless they were with similar-sized animals. Synchronization of activities among group members is important to maintain group cohesiveness (Conradt 1998; Côté et al. 1997b; Ruckstuhl 1999; Ruckstuhl and Neuhaus 2001). If males have shorter foraging bouts than females, mixed-sex groups would be unstable: at some point males would lie down to ruminate, while females would keep grazing and move on. Interspecifically, sexual segregation generally increases with sexual size dimorphism, as predicted by the activity-budget hypothesis (Ruckstuhl and Neuhaus 2000). Intraspecifically, this hypothesis predicts that young adult males should switch to bachelor groups when their body size approaches that of mature males. It also predicts that because they are larger than females but smaller than fully grown adult males, young males should be found in both types of groups, or form groups of their own (Ruckstuhl 1999; Ruckstuhl and Festa-Bianchet 2001). Groups of age-segregated males are sometimes seen in Alpine ibex (Bon et al. 2001), a species with strong sexual dimorphism, and where males take eight to ten years to reach asymptotic body mass. Ibex males aged four to seven years are larger than adult females but smaller than older males (Giacometti et al. 1997). Groups made up only of subadult males have also been reported in Nubian ibex, which tend to form small groups of less than fifteen individuals (Gross et al. 1995). In Alpine ibex, however, males aged two to nine years have very high survival rates (Toïgo et al. 1997), and populations often reach high densities. Consequently, in many ibex populations there are enough young males to form large, age-segregated groups. Groups consisting only of young adult males may be rare in most populations of other mountain ungulates because there may not be enough young males to form a group of sufficient size to obtain the antipredator benefits of gregariousness (Ruckstuhl and Festa-Bianchet 2001). At Caw Ridge there were fewer than five four-year-old males in most years.

If young males move to bachelor groups because they outgrow the body size where they can easily synchronize their activity budget within nursery groups, then heavy young males should spend more time in bachelor groups than light males of the same age. We therefore examined the relationship between male mass and group choice at three and four years of age. Correlations between individual mass and the proportion of time spent in nursery herds were very weak and not significant.[1] The propensity to switch from nursery to bachelor groups does not appear affected by the mass of individual young males.

The behavior of young adult mountain goats provided limited support for the activity budget hypothesis (Ruckstuhl 1998) for sexual segregation. Although young males switched from nursery to bachelor groups at the age when they outgrew adult females, larger young males did not switch before smaller males. Adult females in midsummer weighed on average 71 kg, a mass that most males exceeded at four years of age (chapter 6). Males aged six years and older, however, were about 30% heavier than four-year-old males. That large difference may explain why most males aged four years switched back and forth across group types: they may have been unable to satisfactorily synchronize their activity within either nursery or bachelor groups and they were not numerous enough to form groups on their own. A similar behavior was displayed by young bighorn rams, which were intermediate in size between mature rams and adult ewes (Ruckstuhl and Festa-Bianchet 2001). Based on relative differences in body mass, one would have expected four-year-old males to spend more time with nursery groups than with bachelor groups. On average, four-year-old males were 10% heavier than adult females and 30% lighter than mature males. The observation that four year olds spent 61% of their time in bachelor groups (fig. 5.2) suggests that factors other than body size contributed to group choice. Differences in time budgets between young males and adult females may have been exacerbated by the increased time spent foraging by lactating females, as reported in several ungulate species (Ruckstuhl and Neuhaus 2002). Alternatively, young males may benefit from joining bachelor groups by establishing a social hierarchy that could affect their future mating success.

Data on time budgets of adult males are required to further investigate sexual segregation in mountain goats. Because there are substantial age-related differences in mass among males aged four years and older, it may also be more difficult for bachelor groups to synchronize activities, compared to adult females in nursery groups. Among bighorn rams, time budgets vary according to individual mass, with large rams spending less time feeding than small ones (Pelletier and Festa-Bianchet 2004). If differences in mass among group members lead to difficulties in maintaining synchronization, bachelor groups may be more unstable than nursery groups.

Individual differences in temperament (Réale et al. 2000), social status, or other characteristics may affect each young male's propensity to remain in nursery groups or join bachelor groups. To test for consistent individual differences in behavior, we examined whether the proportions of time a male spent in nursery herds in consecutive years were correlated. The results were inconclusive. Young males that spent much time in nursery groups one year were often seen in nursery groups the following year, but the correlations were not significant.[2] Although it is likely that a larger sample size would result in significant correlations, there appears to be substantial variability in how individual animals behave from one year to the next.

Overall, our data on the ontogeny of sexual segregation in mountain goats are most consistent with the hypothesis that segregation is a function of differences in time budgets induced by differences in body size, but other factors, such as the development of social relationships and differing risks of predation (Bleich et al. 1997; Bonenfant et al. 2004), may be involved. The striking differences in summer distribution (fig. 4.2) suggest that variables other than differences in time budget may also be involved in determining sexual segregation in this population. As we suggested in the previous chapter, males may be either less susceptible to or more tolerant of the risk of predation than females. Data on activity budgets of different sex and age classes are essential to further elucidate the reasons for sexual segregation in this species. Ruminants alternate long bouts of feeding with long bouts of ruminating or resting; therefore, a reliable estimate of their time budget requires continuous records of activity of known individuals. Recent claims to refute the activity budgets hypothesis that do not provide extensive data on activity budgets are of limited value (Bowyer 2005; Mooring et al. 2003).

Postweaning Associations of Mothers and Offspring

In many ungulates, the mother–offspring association ends either at weaning or soon before the birth of the next offspring. In some species, however, mother–offspring associations can persist beyond the first year of life, particularly for daughters (Clutton-Brock et al. 1982; Green et al. 1989). Earlier research on mountain goats suggested that mother–yearling associations were common, although apparently yearlings did not obtain milk from their mothers (Hutchins 1984). At Caw Ridge, many yearling goats were closely associated with their mother (fig. 5.3). Some yearlings appeared to obtain milk during short suckles, as milk was seen dripping from their mouths. A few two year olds and the exceptional three year old also remained in close association with their mother. Behavioral associations between a mother and a yearling or older offspring were

Figure 5.3. A mother with her kid and associated yearling. Photo by F. Pelletier.

obvious. The offspring followed closely behind its mother, often foraged next to her, and when resting mother and offspring were very close or in physical contact. In eighteen cases where the associated juvenile had been marked as a kid (and therefore its mother had been identified), juveniles associated only with their mother. Females did not tolerate close proximity of other juveniles and threatened them with their horns. Therefore, we are confident that all associations between an adult female and a juvenile were cases of protracted maternal care.

In some ungulates, extended maternal care is associated with poor environmental conditions. In bighorn sheep, for example, there is generally a very weak mother–yearling association (Festa-Bianchet 1991), but at high population density many yearlings associate with their mother, possibly because extended maternal care becomes more important to ensure offspring survival when resources are scarce (L'Heureux et al. 1995). At Caw Ridge, almost all yearlings remained with their mother until a few days before the birth of new kids in late May. Some mothers tolerated the proximity of their yearling during parturition and many yearlings remained with their mothers throughout the summer. A few yearlings were still with their mother in early summer but became independent as the summer progressed.

We classified yearlings and older offspring as associated with their mother if they remained with her until at least July 15, but most of these associations persisted until September. To examine the effects of offspring sex on the probability of association, we first considered the behavior of all yearlings and two year olds that survived to midsummer (table 5.1). Among yearlings, sex had no effect on the probability of asso-

TABLE 5.1
Associations between Mountain Goat Mothers and Their Offspring, 1991–2003.

Offspring			Associated			
Age	Sex	Yes	No	% Yes	G	P
Yearling	Male	53	49	52.0	0.43	0.51
	Female	51	39	56.7		
2 years	Male	2	67	2.9	11.06	0.0009
	Female	14	51	21.5		

For offspring older than one year at Caw Ridge, Alberta. Only associations that lasted until at least July 15 were considered. G tests and P values refer to comparisons in the probability of association according to sex, within each age category.

ciation. Among two year olds, very few males were associated with their mother, but one in five females was associated. All but one of the associated two year olds had been associated as a yearling. All three-year-old males were independent of their mothers. Only two of seventy-five marked three-year-old females remained with their mother.

We had a smaller sample to examine the effects of mother's age and presence of a kid of the year on the probability of association, because we could only include yearlings whose mother was known. Because we often did not know the mother of unmarked yearlings that did not associate, this sample was biased in favor of associated yearlings. Age of the mother had no effect on the probability of association for either yearlings or two year olds (table 5.2). Almost all yearlings whose mother was not nursing a kid the following year remained associated, while mothers nursing a new kid were unlikely to associate with their previous offspring (table 5.3).

We could not detect any benefits of protracted association on either juvenile survival or mass gain. Mass gain from yearling to two year old was greater for independent than for associated yearlings,[3] suggesting that yearlings may have been less likely to associate with their mother if they were already in good condition. Protracted maternal care did not have a positive influence on yearling survival or physical development and was most commonly seen for mothers that were not nursing a new kid. Perhaps yearling–mother associations occurred mostly for relatively poor mothers that were unable to care for a new kid and therefore continued to care for their yearling. Consequently, it is possible that had the associated yearlings not received protracted care, they would have been

TABLE 5.2

Associations between Mountain Goat Mothers of Different Ages and Their Offspring, 1991–2003

Age		Associated				
Offspring	Mother	Yes	No	%Yes	G	P
Yearling	4–7	34	15	69.4	0.04	0.98
	8–10	35	15	70.0		
	11+	32	15	68.1		
2 years	4–7	6	11	35.3	1.67	0.43
	8–10	4	19	17.4		
	11+	5	15	25.0		

For offspring older than one year at Caw Ridge, Alberta. Only associations that lasted until at least July 15 were considered. G tests and P values refer to comparisons in the probability of association according to maternal age, within each category of offspring age.

TABLE 5.3
Associations between Mountain Goat Mothers with and without a Kid of the Year and Their Offspring, 1991–2003

Offspring age	Kid present	Juvenile associated				
		Yes	No	% Yes	G	P
Yearling	Yes	59	43	57.8	19.79	0.0001
	No	45	3	93.7		
2 years	Yes	11	42	20.8	2.96	0.08
	No	5	6	45.5		

For offspring older than one year at Caw Ridge, Alberta. Only associations that lasted until at least July 15 were considered. G tests and P values refer to comparisons in the probability of association according to whether or not the mother was accompanied by a kid, within each category of offspring age.

smaller or less likely to survive than independent yearlings, in which case protracted maternal care may be considered compensatory.

The Social System of Female Mountain Goats: An Antipredator Adaptation of Individuals

The social system of mountain goat females during the summer is similar to that of bighorn sheep and different from that of many cervids. There are no behavioral associations among related adults and no spatial substructuring of the population. While many deer form subgroups with distinct home ranges and often consisting of matrilineal family units (Albon et al. 1992), mountain goats and bighorn sheep (Festa-Bianchet 1991) associate with all other females in the population and do not preferentially seek out relatives to form subgroups.

Social organization depends upon the behavior of many individuals, responding to their own needs but also to the behavior of conspecifics (Clutton-Brock 1989b). Consequently, social organization is not an adaptation and cannot evolve through natural selection. It should be expected to change rapidly over both space and time. In very small populations there may be more solitary individuals, and the degree of sexual segregation may vary. Propensity to associate with other individuals under different circumstances, however, is an individual trait, subject to selection that could lead to adaptation. For mountain goats, the main benefit of gregariousness is likely increased protection against predation, as suggested for other herbivores (Illius and Fitzgibbon 1994; Risenhoover and Bailey 1985a). Mountain goats in patchy habitat in rugged and

BOX 5.2
Behavioral Associations among Adult Kin

Although kin groups appear to be the basic social unit of many ungulates (Albon et al. 1992; Clutton-Brock et al. 1982; Nelson 1998; Ozoga and Verme 1984), few studies have performed demographic analyses to estimate the likelihood of co-occurrence of adult relatives, a basic necessity for kin grouping (King and Murie 1985). Adult relatives that coexist may associate with each other, but the importance of kin groups in the social organization of gregarious ungulates should vary with the extent of temporal overlap of female kin. If related individuals are unlikely to overlap in time, they cannot form social units.

We did not know the mothers of many goats that were caught after weaning, therefore we could not estimate matriline size or examine the probability of coexistence of sisters or grandmother–granddaughter pairs. Instead, for females with known mothers, we calculated the proportion aged three years or older for which the mother was still alive. We defined "adult" females as those aged at least three years, the minimum age of primiparity during our study. The youngest age at which we documented a female with an adult daughter was seven years. As expected, the probability of coexistence declined with female age, but the mothers of most females aged three to six years were alive (fig. 5.4). Therefore, most adult females had at least one closely related adult female with whom they could associate. The only other published analysis on temporal overlap of uterine kin in ungulates is for bighorn sheep (Festa-Bianchet 1991), where the mothers of about 64% of adult ewes aged five or six years were still alive. We obtained very similar results for mountain goats: 62% of females aged five or six years coexisted with their mother. In mountain goats, social behavior may play an important role in female reproductive success (Côté 1999), therefore one could expect behavioral associations between females and their adult daughters. Such associations may help win aggressive interactions, although female kin coalitions have not been reported in any ungulate.

To examine whether or not related adult females were likely to be in the same group, we used the method described by Festa-Bianchet (1991), which first estimates the probability P_t that a female would be in the same group as her daughter if they were distributed at random among groups:

$$P_t = (n - 1)/(N - 1)$$

where N is the total number of females aged three years and older in the population and n is the number in the group. The expected number of times that a mother and her daughter should be seen together if they chose groups independently (E_t) is therefore

---

BOX 5.2
Continued

$$E_t = \sum_{i=1}^{t} \frac{n_j - 1}{N - 1}$$

where t is the number of valid sightings of the daughter during the summer. Valid sightings were those when at least five days had elapsed since the last sighting used, to minimize dependence among observations. We had a mean of 12.4 (± 0.4 SD) valid sightings per mother–daughter pair per year.

The number of valid sightings when mother and offspring occurred together was compared to the expected number E_t with Wilcoxon matched-pairs tests. We found little evidence that adult female kin associated with each other. Mother–daughter pairs were in the same group in 50% of cases, only 5% more than expected had they been distributed at random with respect to each other (Wilcoxon matched-pairs test, $Z = -1.637$, $P = 0.10$, $n = 77$ mother–daughter pairs). The expected proportion of co-occurrences (45%) is high because many of the groups seen in midsummer included almost all adult females, and related females were necessarily together. Similarly to bighorn sheep ewes (Festa-Bianchet 1991), mountain goat females are highly philopatric but do not preferentially associate with female relatives.

---

mountainous terrain should form smaller groups than on Caw Ridge because they may not require the level of antipredator protection that a large group can provide in open habitat. Where goats use very rugged and inaccessible terrain, the risk of predation may be lower than in the open, rolling terrain of Caw Ridge. In addition, small patches of habitat may not have sufficient forage to sustain large groups.

Other possible benefits of gregariousness could include learning the location of the best seasonal foraging areas, travel routes, escape terrain, and shelters from inclement weather. Given the protracted associations of mothers and offspring and the small area of Caw Ridge, however, it seems likely that much of that spatial information could be learned during the period of maternal care. Consequently, the need to learn from conspecifics is unlikely to explain the very large groups formed by adult females and young during summer. Another potential benefit of gregariousness is group defense, but all interactions we saw between goats and predators involved single individuals. Group defense was never observed (Côté et al. 1997a).

If the main benefit of forming a group is to increase the probability of predator detection, there may be little incentive to associate with one's relatives. A measurable behavioral benefit of gregariousness is a decline in individual vigilance and an increase in the time spent foraging, because larger groups are more efficient at detecting approaching predators. In mountain goats, individual vigilance continues to decrease as group size increases to at least ten to fifteen individuals (Haviernick 1996; Risenhoover and Bailey 1985b). We were unable to estimate matriline size, but it is likely that few adult females had more than three to four close relatives in the population (fig. 5.4), and therefore families could not form the large groups typical of ungulates in open habitats. In bighorn sheep, which also do not form kin groups but instead associate with any female from their population, the average size of matrilines (defined as coexisting adult females sharing a female ancestor within two generations) was only approximately 1.8, and one quarter of adult ewes had no known living female uterine relatives (Festa-Bianchet 1991). Lack of kin grouping has also been reported in free-ranging domestic sheep (Lawrence 1990; Lawrence and Wood-Gush 1988) in open habitats. In more fragmented goat populations, small groups of related females restricted to narrow cliff bands may form permanent associations, simply through phylopatry. At Caw Ridge in summer, forage was distributed over a wide area, steep escape terrain was limited, and goats relied mostly on detecting predators at a distance to avoid predation.

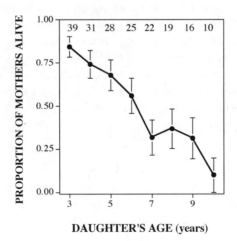

DAUGHTER'S AGE (years)

Figure 5.4. Proportion (± SE) of adult female mountain goats aged three to ten years born in 1989 to 1999 that coexisted with their mother. Sample sizes are indicated along the top of the graph.

Implications of the Social Organization of Mountain Goats

The social organization of mountain goats has several important implications for behavior, population dynamics, and conservation, which will be considered in greater detail in later chapters. First, all female goats had regular opportunities to interact with all other females in the population. Therefore, they could form dyadic relationships based on individual recognition and on the outcome of previous social interactions (Côté 1999). The establishment of dyadic relationships meant that when two goats came within a short distance, the subordinate individual knew that it should avoid the dominant, reducing the risk of injury through aggression. Second, lack of spatial structuring in population dispersion should lead to lack of spatial structuring in population dynamics because all members of the population face the same levels of resource availability and predation risk. Differences in individual performance, therefore, should be due to individual differences in the ability to acquire or utilize resources and not to differences in the quality of habitat used by different goats. Third, as has been shown for bighorn sheep (Ross et al. 1997), the antipredator strategy adopted by mountain goats on Caw Ridge is likely to be effective against cursorial predators that can be detected from a distance, such as wolves, but may not be very effective against stalking predators such as cougars, particularly individual predators that specialize in hunting in the type of terrain used by mountain goats. The behavior of adult males, however, suggests that they could be more vulnerable to predation than adult females. Finally, mountain goat harvest strategies that aim to remove a certain proportion of animals from a population should consider habitat differences. In large areas of open habitat where goats form large groups, harvests would likely affect a large population. In a more patchy habitat, where goats may be restricted to small areas of suitable habitat, harvests may have a devastating impact on some small subgroups but leave neighboring groups unaffected.

Summary

- Mountain goats were sexually segregated during summer and formed two types of group: nursery groups of females, kids, yearlings, and young males and bachelor groups of adult males, mostly aged four years and older.
- Adult females were highly gregarious and were rarely seen alone. Adult males were less gregarious than adult females and 10% of

adult male sightings were of lone males. From May to August, nursery group size increased, whereas bachelor group size decreased.

- Males switched gradually from nursery to bachelor groups at three and four years of age. They were not expelled from nursery groups by adult females, and they won most social interactions with adult females. Individual mass of young males did not affect the timing of their switch from nursery to bachelor groups. Differences in time budgets between adult males and females may explain sexual segregation in mountain goats, although differences in predation risk are also likely to play a role.
- Approximately 54% of yearlings of both sexes and 21% of two-year-old females were closely associated with their mother. Post-weaning association was independent of maternal age or offspring mass but was more frequent when the mother did not have a kid of the year. There were no measurable consequences of protracted maternal care on yearling survival, horn growth, or mass gain.
- Although most adult mountain goat females had at least one adult relative in the population, females did not preferentially associate with relatives.
- The social system of mountain goats during summer appears to be mostly an adaptation to avoid cursorial predators.

Statistical Notes

1. Spearman correlations between the proportion of time spent in nursery herds by males aged three and four years and their mass on July 15 at two and three years of age. Data from 1995 to 2000.

Age (years)	Mass at two years			Mass at three years		
	R	P	N	R	P	N
3	0.182	0.55	13	−0.117	0.76	9
4	−0.108	0.80	8	0.048	0.91	8

2. Spearman rank correlations between the proportion of time a male spent in nursery herds as a three and as a four year old, $r_s = 0.46$, $n = 14$, $P = 0.10$; as a four and a five year old, $r_s = 0.49$, $n = 11$, $P = 0.12$.

3. For marked yearlings, survival to two years was 82% for sixty-two associated ones and 87% for twenty-three independent ones ($G = 0.28$,

P = 0.60). There were no differences in mass among associated and independent yearlings as kids, yearlings, or two year olds (all P's > 0.3, accounting for sex of yearlings and two year olds). Mass gain from kid to yearling age was not affected by protracted association ($F_{1,17}$ = 0.71, P = 0.41 accounting for sex), and mass gain from yearling to two year old was greater for independent yearlings (males: 18.2 ± 2.55 kg [n = 3]; females: 16.2 ± 2.58 kg [n = 5]) than for associated ones (males: 14.5 ± 3.25 kg [n = 10]; females: 10.8 ± 4.25 kg [n = 13]) ($F_{1,28}$ = 10.33, P = 0.003 after accounting for sex).

CHAPTER 6

Body and Horn Growth

The importance of body mass in ecology and life-history evolution cannot be understated (Peters 1983). Much research has been carried out on the effects of interspecific differences in body size, and in recent years increasing attention has been paid to the effect of variability in body size within a population (Bérubé et al. 1999; Gaillard et al. 1997, 2000b; McElligott et al. 2001). These studies suggest that large individuals generally perform better than small ones. Here we examine the physical development of mountain goats, in particular body mass and horn size. We show that sexual dimorphism in body mass is much greater than in horn size, and age-specific patterns of horn growth of males and females are substantially different. Sexual dimorphism develops almost entirely after weaning, and dimorphism in horn size develops faster than dimorphism in body mass. Both sexes reach asymptotic mass after several years of growth. These findings have substantial implications for social behavior, life-history strategy, population dynamics, and management of mountain goats.

Because large mammals are usually difficult to capture, very few studies have obtained repeated mass or body measurements of marked individuals. The inability to monitor individual body mass is a serious limitation of much large-mammal research, in view of the wealth of information that can be extracted from a knowledge of individual mass and of seasonal and multiyear changes in mass. For examples, studies of bighorn sheep have revealed that individual mass affects survival of lambs and of old ewes (Festa-Bianchet et al. 1997), longevity of adult ewes (Bérubé et al. 1999), and the fitness costs of reproduction (Festa-Bianchet et al. 1998).

Age- and sex-specific changes in both mass and horn size are also a useful indicator of resource availability and provide important clues to the evolution of life histories for both males and females (Festa-Bianchet et al. 2004; Festa-Bianchet and Jorgenson 1998; Jorgenson et al. 1998; Leblanc et al. 2001). Recently, individual mass and horn size have been related to male reproductive success in bighorn sheep and fallow deer (Coltman et al. 2002; McElligott et al. 2001).

Because the sex- and age-specific body and horn growth of mountain goats plays an important role in many aspects of their ecology discussed in later chapters, we describe those patterns in some detail. These include the surprisingly slow, multiyear increase in mass for both sexes, the rapid horn growth of young males, and the similarity in asymptotic horn length between the sexes.

Most goats were only caught two or three times in their lifetime, typically first as a kid, yearling, or two year old, and then again at an older age to fit a visual collar for females, or to collect more measurements for males. In recent years, we obtained repeated measures of body mass with electronic platform scales (Bassano et al. 2003). Between 1988 and 2003, we weighed captured goats on 508 occasions. Between 2001 and 2003 we obtained an additional 416 mass measurements using platform scales. Males aged four years and older seldom came to the trap site, and we obtained only forty-nine mass measurements of adult males, all but one before July 15.

How Heavy Are Mountain Goats? Effects of Age, Sex, and Season

Goats of all sex–age classes appeared to gain mass from early June to mid-September. Goats were more attracted to the salt bait earlier in the season, so that few mass measurements were obtained after July. We do not know when goats stopped accumulating mass at the end of summer. We expected that the rate of mass gain would decrease over summer, as the quality of available vegetation declined and as goats, particularly adults, restored lost fat reserves and muscle mass, similar to the pattern observed for bighorn sheep at Ram Mountain (Festa-Bianchet et al. 1996). A decreasing rate of mass gain from early to late summer should be better described by a quadratic than a linear relationship. In no case, however, did a quadratic regression improve the fit over a linear relationship: goats of all sex and age classes appeared to gain mass linearly from June to September. The quadratic term, however, was nearly significant for females older than five years (P = 0.08) and, inevitably, mass gain must slow down before winter.

It is possible that mass gain was nearly linear for much of the summer because goats had access to high-quality forage for a longer period than bighorn sheep at Ram Mountain, as suggested by differences in the seasonal changes of fecal crude protein in the two study areas (fig. 4.4). Mass gain rate by adult goats probably began to decrease sometime in August, but because most mass measurements were obtained earlier in the season, linear regressions provided an adequate fit to the data. We therefore used age- and sex-specific linear regressions to adjust individual mass to mid-summer (July 15) for all sex–age classes.

Little is known about the birthweight of mountain goat kids. Seven captive-born kids weighed an average of 4.1 kg one week after birth (Carl and Robbins 1988). Birthweight for free-ranging mountain goats is likely about 3 kg (Brandborg 1955; Lentfer 1955). Assuming a linear mass gain from birth at 200 g/day (fig. 6.1), then on the median birthdate of May 26 (Côté and Festa-Bianchet 2001a) kids would have weighed about 3.3 kg.

Kids of both sexes gained mass linearly during summer (fig. 6.1). Overall, there was no difference among sexes in mass or in rate of mass gain.[1] Kids averaged about 8.7 kg on June 20, at just under one month of age, and 26 kg by September 15. We have previously shown that fecal crude protein values in early June affect the mass of kids (Côté and Festa-Bianchet 2001a). Reanalysis including more years of data confirmed that

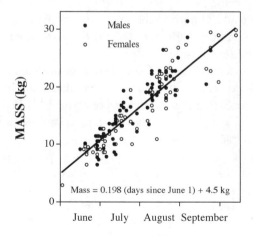

Figure 6.1. Mass of mountain goat kids at Caw Ridge during summer, 1989 to 2003.

June fecal crude protein explained 12% of the variability in kid mass adjusted to July 15. Together, date of capture and fecal crude protein explained 84% of variability in kid mass (n = 113 kids) for years when information on fecal crude protein was available.

Kid mass, however, was affected also by maternal age (chapter 7). For kids with known-age mothers, males were about 6% heavier in midsummer than females (fig. 7.4). Although on average male kids are slightly heavier than female kids,[2] there is much overlap in mass between the sexes (fig. 6.1).

A comparison of body mass of mountain goat kids and bighorn lambs provides an insight into the slow physical development of mountain goats compared to other mountain ungulates. Adult female goats are about 13% heavier than adult bighorn ewes (80 vs 71 kg for mass adjusted to mid-September), but mountain goat kids are smaller than female bighorn sheep lambs. In the Ram Mountain bighorn population at low density, female lambs averaged about 10.2 kg on June 20 and 27.5 kg on September 15, gaining about 200 grams per day, while male lambs were 10% heavier (Festa-Bianchet et al. 1996). Mountain goat kids (sexes combined) were 1.5 kg lighter than female bighorn sheep lambs over the summer, and gained mass at the same rate. By mid-September, female bighorn lambs were 6% heavier than mountain goat kids.

There was little sexual dimorphism in body mass for yearling goats at the beginning of the summer. The intercept of the linear regression of yearling mass on capture date, corresponding to mass on June 1st, was 24.3 kg for females and 24.7 kg for males (fig. 6.2). Therefore, goats were about 1.5 kg (or 6%) lighter on June 1 as yearlings than they were as kids the previous September 15. Before one year of age mountain goats show very little sexual dimorphism in body mass.

Sexual dimorphism began to develop during the second summer of life. Yearling males gained mass 25% faster than females (225 vs 180 g/day),[3] so that by September 15 they were about 12% heavier than females (48.8 vs 43.5 kg) (fig. 6.2). Mass of yearlings was not affected by fecal crude protein in early June.[4]

There was little change in sexual dimorphism over the second winter of life, because by June 1st two-year-old males were 11% heavier than two-year-old females (37.2 vs 33.6 kg). Two-year-old goats of both sexes were 23% lighter on June 1 than on September 15 as yearlings. Actual winter mass loss must have been greater, because yearlings likely gained some mass after September 15 (fig. 6.3), and all goats appeared to be gaining mass by June 1st. During summer two-year-old males gained mass more rapidly than females of the same age,[5] so that by mid-September

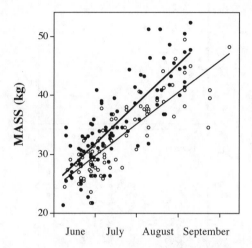

•   Males Y = 0.225 X + 24.7

o   Females Y = 0.180 X + 24.3

Figure 6.2. Mass of male and female mountain goat yearlings at Caw Ridge during summer, 1989 to 2003. Mass gain equations are based on day 0 being June 1st.

males weighed about 70 kg and females about 60 kg, a 17% difference. Mass gain by two year olds appeared linear from early June to late August.

Adult females and adult males gained mass during summer, therefore we adjusted mass of all individuals to July 15 to compare different sex and age classes. Males five years and older averaged 98 kg ± 8.34 SD in mid-July (n = 36), 12% less than the 112 kg reported for nine adult males weighed from late June to mid-July in Olympic National Park, Washington (Houston et al. 1989). On the other hand, adult females on Caw Ridge were 15% heavier than those in Olympic National Park, where females five years of age and older averaged 62 kg in early summer (Houston et al. 1989). The average mass of Caw Ridge adult females (five years and older) on July 15 was 71 kg. Adult sexual dimorphism appears to vary substantially among mountain goat populations: males in Olympic National Park may be as much as 80% heavier than females, while at Caw Ridge males are only about 38% heavier than females. In both populations, however, there was a limited sample size for mature males.

Because adult females weighed about 58 kg on June 1 and 80 kg on September 15 (fig. 6.3), on average they gained and lost a minimum of about 22 kg each year. The average winter loss was at least 27% of

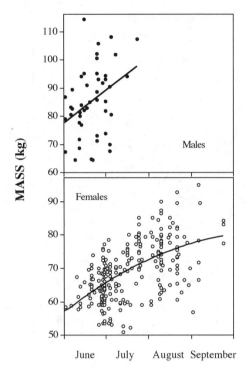

Figure 6.3. Mass of male and female mountain goats aged four years and older at Caw Ridge during summer, 1989 to 2003. No mass measurements of adult males were obtained after July.

late-summer mass, and summer gain was a minimum of 38% of late-spring mass. There was considerable individual variation: late-summer mass ranged from 65 to 95 kg. We could not estimate seasonal mass changes for adult males because we had no weights after mid-July.

Goats of both sexes gained mass with age until about six years, and females appeared to lose some mass after about ten years of age (fig. 6.4). The apparent increase in average mass with age may be misleading if light females had greater mortality than heavy ones. In both bighorn sheep and roe deer, adult female longevity is correlated with mass as a young adult: lighter females die younger, and heavier females live longer but lose mass in their later years (Bérubé et al. 1999; Gaillard et al. 2000b). We did not find a link between adult mass and longevity in mountain goat females,[6] but given the limited sample size this possibility cannot be excluded.

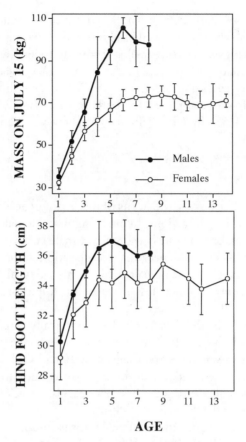

Figure 6.4. Mass and hind foot length (± SD) adjusted to July 15 of mountain goat yearlings and adults at Caw Ridge, 1988 to 2003. The last data point in each series shows the average for all males aged eight years and older, and all females aged fourteen years and older.

Age and Sex Effects on Body Size

Changes in mass could be due to changes in body size but also in fat and protein accumulation. To monitor growth in skeletal size, we compared age- and sex-specific hind foot measurements. During summer, the feet of yearling males grew faster (0.5 mm per day) than those of females (0.3 mm per day). Among two year olds, however, the pattern was reversed: the increase in hind foot length during summer was significant for females (0.2 mm per day) but not for males (0.15 mm per day).[7] For goats aged three years and older, foot length did not increase during summer. Consequently, for comparisons of hind foot length we adjusted measurements for yearlings and two year olds according to age- and sex-specific regressions on date of capture, and used unadjusted measurements for goats aged three years and older. Changes in foot length suggested that skeletal growth was completed for both sexes at four years (fig. 6.4). Therefore, goats continued to gain mass for a few years after completing skeletal growth.

Horn Growth

Like those of other bovids, the horns of mountain goats are mostly used during intraspecific interactions. In most ungulates, females have smaller horns than males, presumably because male–male competition for mates has greater fitness consequences than competition among females, and horns play a prominent role in male–male combat. Sexual dimorphism in horn size is generally much less marked for Rupicaprinae, including mountain goats, than for most other ungulates. For example, in wild sheep and true goats, the horns of males are typically three to four times as long and much thicker than those of females. In contrast, among Rupicaprinae, the horns of males are usually only 10 to 20% longer or thicker than those of females, and sexual dimorphism is due mostly to greater horn growth by males during the first few years of life (Côté et al. 1998b; Miura 1986; Pérez-Barbería et al. 1996).

The functional significance of female horns in ungulates is not well understood (Estes 1991; Packer 1983), and in some species, including mountain goats, horns may also be used for antipredator defense (Côté et al. 1997a; Locati 1990). The stiletto-like horns of mountain goats are sharp, pointed, and dangerous. Female ungulates use their horns in intrasexual aggressive interactions, and the rate of female–female aggression in mountain goats is much higher than in most other ungulates (Fournier and Festa-Bianchet 1995).

If the size of mountain goat horns played an important role in male–male competition for access to estrous females, similarly to the antlers of cervids and the horns of many other bovids, we may expect considerable sexual dimorphism in horn size and a longer period of horn growth for males than for females, as commonly observed in species where males fight by clashing or wrestling with horns or antlers. Mountain goats, however, do not fight through horn contact, but rather attempt to stab or gore each other during escalated aggressive encounters (Côté 2000;

Figure 6.5. Mountain goats have an antiparallel fighting style, the head of each goat facing the rump of the opponent. They circle while fighting and normally arch their shoulders in a present threat. Photo by S. Côté.

Geist 1971), similarly to other rupicaprins (Locati and Lovari 1990). They fight head to rump in a particular antiparallel style (fig. 6.5). Most aggressive encounters among females are ritualized displays, and contact occurs in < 0.1% of interactions (Côté 1999). Aggressive behaviors include present threat (broadside orientation with apparent size enhanced by arching the back), horn threat (display or use of the horns), rush threat (sudden movement toward an antagonist), and orientation threat (a low-intensity form of rush threat involving walking). Submissive behaviors include orientation avoidance (avoiding the opponent by walking or staring away) and rush avoidance (quickly moving away from the antagonist). For these behaviors, large horns may not be very useful, and long horns may be at greater risk of breakage (Alvarez 1994). The frequent aggressive interactions among females make horns useful for both sexes.

In most Caprinae, horn growth continues throughout life. It decreases with advancing age (Alvarez 1990; Jorgenson et al. 1998) and stops during winter in temperate and arctic regions (Bunnell 1978). In chamois and mountain goats, individuals whose horns grow little in their early years show greater horn growth in later years (Côté et al. 1998b; Pérez-Barbería et al. 1996). In females, reproduction reduces horn growth, presumably because resources are allocated to lactation rather than to tissue growth. For Japanese serow, whose horns are similar to those of mountain goats, Miura et al. (1987) found that a female's horns grew less in years when she produced a kid than in years when she did not, and in young bighorn sheep ewes primiparity was associated with a decrease in horn growth (Festa-Bianchet and Jorgenson 1994). Females may face a trade-off between horn growth and reproduction.

When the horns of yearling and adult mountain goats stop growing in winter, they form a distinct ring, called an annulus (fig. 6.6). Horn annuli provide an easy and reliable way of aging adult goats, although the age of older individuals is occasionally underestimated (Stevens and Houston 1989). In some years the horns of adults older than six years do not form a distinct annulus. The horns of kids grow through the winter, although the rate of growth slows considerably. The first distinct annulus is formed during the second winter, at about 1.5 years of age; thereafter, annuli form in early winter.

We used a measuring tape to record total horn length, length of each annulus along the outside curve, basal circumference of each horn, and the circumference of each annulus to the nearest mm. For goats aged eight years or older, we measured only the first seven annuli because later ones were often indistinct (Brandborg 1955; Côté 1999; Stevens and Houston 1989). There was no directional asymmetry in either total horn

Figure 6.6. Measuring horn length on #206 when he was seven years old. Note the clear annuli. The first annulus is formed at 1.5 years of age in mountain goats. Photo by S. Hamel.

length or base circumference (Côté and Festa-Bianchet 2001b), indicating that horns were symmetrical.

Horns became visible on kids by July. By September both sexes had soft, blunt horns about 3 cm long. Horn length more than tripled between five and twelve months of age, so that by June 1st the horns of yearlings averaged 11.4 cm for males and 10.2 cm for females. The horns of yearlings grew linearly and rapidly from June to September, with males gaining on average 0.44 mm/day and females 0.32 mm/day. Therefore, we adjusted horn length of yearlings to mid-July for further comparisons. Yearling goats had slightly longer horns in years when fecal crude protein content in early June was high,[8] suggesting that spring nutrition affected horn growth.

For both sexes, horn length of yearlings was positively correlated with body mass,[9] but even when the effects of sex and mass were considered, horn length increased with early-June fecal crude protein.[10] For a given body mass, yearlings grew longer horns in years when high-quality forage was available early in the season. That may suggest that yearlings allocated more resources to body rather than to horn growth when resources were scarce, as reported in bighorn rams (Festa-Bianchet et al. 2004), but the weakness of the relationship makes it difficult to interpret.

Horn growth during summer by two year olds varied according to sex: the horns of males only grew by 1 mm every five days, those of females grew more than twice as fast, averaging 1 mm every two days. Horn base circumference, on the other hand, grew by 1 mm every ten days for males, but only by 1 mm every sixteen days for females. Between two and three years of age sexual dimorphism in horn length decreased, while dimorphism in base circumference increased (fig. 6.7).

By three years of age, horn growth during summer was so slow that the relationship between horn length and capture date was not significant for either sex.[11] Therefore, for all goats aged three years and older we did not adjust horn measurements according to capture date.

Most horn growth took place during the first three years of life: the average horn length of three-year-old females was 86% of the length at seven years of age and older, while the horns of three-year-old males were about 94% as long as those of males aged five years and older. Horn base circumference at three years was about 93% of the asymptotic value for both sexes (fig. 6.7). The amount of horn grown during the first two years of life (to the first annulus) accounted for a greater proportion of adult horn length for males, where it explained 55% of the variability in horn length at age four (n = 15), than for females, where it explained 22% of variation at four years (n = 30).

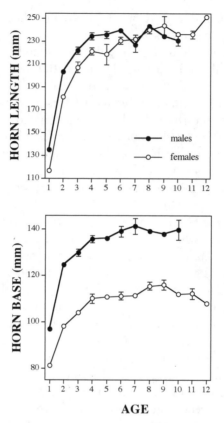

Figure 6.7. Length and base circumference of mountain goat horns by sex and age (mean ± SE). Measurements for yearlings and two-year-old goats were adjusted to July 15, all other measurements were not seasonally adjusted. Sample size for males declined from 74 yearlings to 5 seven year olds and was 1, 1, and 3 for males aged 8, 9, and 10 years. For females, sample size declined from 67 yearlings to 4 9 year olds and was 1, 4, and 1 for females aged 10, 11, and 12 years.

Males had much longer first annuli than females, whereas females grew longer annuli than males in most years from two years of age onward (fig. 6.8).[12] Consequently, the horns of males were longer than those of females until about six years of age. For goats seven years of age and older horn length averaged 23.5 cm ± 0.12 SD for both sexes. The sixteen goats with the longest horns included eight females and eight males. The longest horn belonged to a nine-year-old female (26.9 cm), the longest-horned male (25.7 cm) was a five year old. Males had thicker horns than females at all ages (fig. 6.7). The thickest horn base (circum-

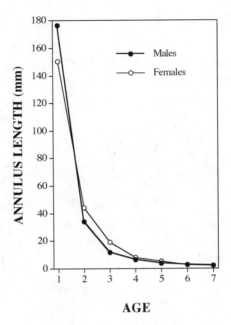

Figure 6.8. Length of the first seven horn annuli of mountain goats of both sexes. The first annulus includes growth as a kid and as a yearling, because the kid growth ring is indistinct. Sample sizes for males averaged 34 and decreased from 101 yearlings to 6 seven year olds. Sample sizes for females averaged 42 and decreased from 98 yearlings to 13 seven year olds. Standard errors were too small for the figure's resolution.

ference 15.2 cm) belonged to a seven-year-old male. Long horns are possibly weaker and thus easier to break, and may incur higher thermoregulatory costs (Picard et al. 1994). The thicker, but not longer, horns of mature male mountain goats compared with those of mature females suggest that stronger horns, resistant to breakage when stabbing, may be the most useful weapon shape for the species' fighting style.

Both sexes showed compensatory horn growth (Côté et al. 1998b). Goats that grew long first annuli tended to grow less horn over the following two years, although the negative relationship between early and later horn growth was weaker for males than for females (fig. 6.9). Compensatory horn growth has also been reported in chamois (Pérez-Barbería et al. 1996) and in Dall's sheep rams (Bunnell 1978; Hoefs and Nowlan 1997). In bighorn rams, however, Festa-Bianchet et al. (2004) found no evidence of compensatory horn growth. In Spanish ibex males, early and late horn growth were positively correlated: rather than

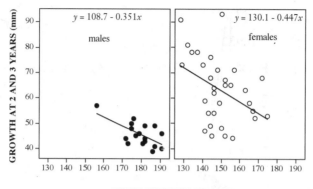

FIRST ANNULUS LENGTH (mm)

Figure 6.9. Compensatory growth in horn length of mountain goats. For both sexes, the sum of the second and third annuli was negatively correlated with the length of the first annulus (males: r^2 = 0.40, n = 16, p = 0.006; females r^2 = 0.168, n = 31, P = 0.02). The results for males were heavily influenced by one individual with a very short first annulus. If that male was excluded, the relationship was not significant (r^2 = 0.12, n = 15, P = 0.19).

showing compensatory growth, males that grew more horn than average when they were young also grew larger horns in later years (Fandos 1995). In Alpine ibex, only females showed compensatory growth (Toïgo et al. 1999). These interspecific comparisons suggest that the advantage of having long horns is greater for males in wild sheep and true goats than for males of mountain goats and probably also other rupicaprids. A limited role of horn size in male–male competition in mountain goats is suggested by the lack of sexual dimorphism in horn length despite the strong sexual dimorphism in body mass and the polygynous mating system. In species that fight by clashing horns, such as bighorn sheep and probably ibex, horn size is important in determining male mating success (Coltman et al. 2002). In those species, one would not expect compensatory horn growth in males because there may be no advantage for males that have grown large horns in early life to limit their horn growth in later years. The compensatory growth in horn length of mountain goats suggests that having longer horns relative to other members of the population may not be very important. In addition, the coefficient of variation in horn length for mountain goats males aged four years and older was only 5.1%, compared to 9.7% for the horns of bighorn rams aged six years and older at Ram Mountain (Festa-Bianchet unpublished data). Consequently, one should not expect adult horn length to be strongly correlated with the outcome of agonistic competition or with male re-

BOX 6.1
Horn Tip Wear and the Effects of Lactation on Horn Growth

If goats frequently used their horns for social interactions or for any activities, the horn tips should wear off. We estimated horn-tip wear by comparing the length of the first annulus measured during captures of the same goat in different years. On average, the first annulus was 2.6 mm or 1.6% shorter when remeasured (t_{36} = 4.39, P < 0.001). Neither goat sex nor the number of years between repeated measurements (average 2.7 years, range 1–8) had a significant effect on tip wear. Tip wear likely explains why male horns reach an apparent asymptote in length at five to six years even though they are still growing 1 to 2 mm a year by producing new annuli (figs. 6.7 and 6.8).

In bighorn ewes, those that grew longer horns during the first two years of life tended to have an earlier age of primiparity, presumably because they were generally in better body condition than young ewes that grew short horns (Festa-Bianchet and Jorgenson 1994). Early horn growth in female mountain goats, however, was not associated with variability in age of primiparity. Females that produced their first offspring at three or four years of age did not grow longer horns by three years of age ($\bar{X}$ = 212 mm ± 11.0 SD, N = 10) than females that did not reproduce until they were five years or older (210 mm ± 13.2, N = 8, t = 0.43, P = 0.67). To assess the effects of lactation on horn growth, we compared annuli of four- and five-year-old females that were and were not lactating (Côté et al. 1998b). Lactating four year olds grew less than half as much horn as nonlactating ones (4.8 vs. 10.9 mm, t_{17} = 2.88, P = 0.01). The same trend existed for five year olds (4.4 vs. 5.3 mm) but it was not significant (P = 0.21). Therefore, reproduction by young adult females involved a trade-off with horn growth, supporting the results obtained for Japanese serow (Miura et al. 1987) and bighorn sheep (Festa-Bianchet and Jorgenson 1994).

productive success. Indeed, once age was accounted for, horn length did not predict social rank in adult females (Côté 2000). It seems that for a male mountain goat it is important to have horns, but not necessarily to have longer horns than those of potential competitors.

It has been suggested that horns and antlers of some ungulates may play a role as display organs (Geist 1987), although empirical evidence in support of this idea is meager (Clutton-Brock 1982; Clutton-Brock and McComb 1993). The limited individual variability in horn size, and the presence of compensatory growth suggest that horn size does not play a prominent role in the social displays of mountain goats. Postures apparently aimed at intimidating opponents tend to emphasize body size rather than horn size (Geist 1971).

The Development of Sexual Dimorphism: Telling the Sheep from the Goats

The evolutionary causes of sexual dimorphism have been the subject of considerable interest (Karubian and Swaddle 2001; Linklater 2000; Loison et al. 1999b). It is generally believed that in polygynous species males are selected to develop a larger size than females because they must compete with each other to obtain access to estrous females; therefore, larger males should be dominant and have greater reproductive success than smaller ones. Sexual dimorphism in body mass may have important ecological consequences because the faster growth and larger adult size of males may also be associated with greater mortality, although the association between male-biased mortality and sexual dimorphism in body size is not always strong (Toïgo and Gaillard 2003). Nevertheless, the development of sexual dimorphism as the animal ages may provide clues to male mating strategies and be related to sexual differences in mortality.

We compared the age- and sex-specific mass changes of mountain goats on Caw Ridge and bighorn sheep on Ram Mountain, Alberta, between 1973 and 1985. Ram Mountain has been the site of a long-term study of bighorn sheep ecology since 1972 (Jorgenson et al. 1993b), and from 1973 to 1985 sheep growth was not affected by changes in density (Leblanc et al. 2001). We adjusted the mass of both species to July 15. Bighorn sheep and mountain goats show several striking differences in mass gain patterns: for both sexes, juvenile sheep (aged zero to two years) are heavier than juvenile goats of the same age, but full-grown mountain goat males are as heavy as bighorn rams, and nannies are about 12% heavier than bighorn ewes. The reason for these age-related differences is that bighorn sheep reach asymptotic mass much faster than mountain goats. At three years of age, bighorn ewes averaged about 84% of the average mass of nine-year-old ewes, while rams are about 78% as heavy as rams aged six to eight years. The corresponding figures for mountain goat females and males of the same ages are only 77% and 62% (fig. 6.10). Therefore, mountain goats grow more slowly than bighorns over the first four years of life, although females of both species reach asymptotic mass at about the same age. We have insufficient mass measurements of mature mountain goat males to adequately describe age-related changes in mass after five years of age, but the available data suggest that, like bighorn males, they may achieve asymptotic mass between six and eight years of age.

Sexual dimorphism in body mass also develops at a younger age for bighorn sheep than for mountain goats. At three years of age, male goats

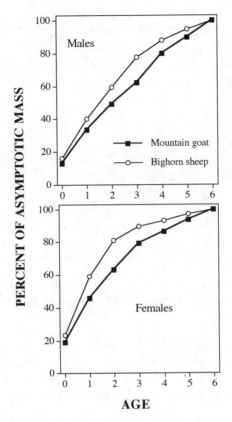

Figure 6.10. The proportion of mass as a six year old reached by male and female mountain goats and bighorn sheep of different ages. Mountain goat data are from Caw Ridge (1988 to 2003), bighorn sheep from Ram Mountain (1973 to 1985, low-density years only).

are only about 16% heavier than females, while male bighorn sheep are 39% heavier than females. Adult dimorphism, as mentioned above, may vary among populations. Our comparisons are mostly between goats on Caw Ridge and bighorn sheep on Ram Mountain. For animals aged six years and older, dimorphism is slightly greater for bighorn sheep than for mountain goats (fig. 6.11). Bighorn rams are about 1.6 times as heavy as bighorn ewes, while mountain goat males are about 1.5 times as heavy as females. The two species, however, differ widely in horn size dimorphism: while mountain goats show no sexual dimorphism in horn length and about 20% dimorphism in horn base circumference, bighorn rams have horns that are about three times both longer and thicker than

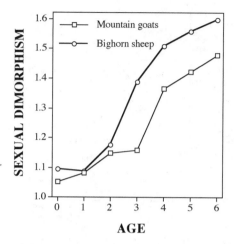

Figure 6.11. Sexual dimorphism (average male mass divided by average female mass, all mass measurements adjusted to July 15) for bighorn sheep at Ram Mountain (1973–1985) and mountain goats at Caw Ridge (1988–2003). Males aged six to eight years are pooled in the "six year old" age class for both species.

those of bighorn ewes. In addition, horn length dimorphism in bighorn sheep increases with age, while in mountain goats it peaks among year-lings and two year olds and declines thereafter. Clearly, horns must play very different roles in the mating system and the social organization of these two species.

Does High Density Slow the Growth of Caw Ridge Goats?

The slow rate of mass accumulation exhibited by young mountain goats on Caw Ridge could be an inherent characteristic of this population, due to the local environmental conditions. The late green-up and harsh cli-mate may select a growth strategy whereby individual goats require many years to reach their asymptotic mass. Alternatively, the slow rate of mass gain could be a constraint due to intraspecific competition. If mass gain were density-dependent, however, it should decline with increasing pop-ulation size, as has been reported in several other ungulate populations (Clutton-Brock et al. 1996; Leberg and Smith 1993; Pettorelli et al. 2002). Most other studies of ungulates also report a greater negative ef-fect of density on growth of males than of females (Leberg and Smith 1993; Leblanc et al. 2001; Toïgo et al. 1999).

During our study, the total number of mountain goats in June varied from 81 to 152, including 35 to 59 adult females. Therefore, the popula-

---

BOX 6.2
Measuring Population Density

For mountain goats on Caw Ridge, density and population size are approximately equivalent, as all females and young goats shared a common home range, whose extent apparently remained the same year after year regardless of population size. Similarly to bighorn sheep, mountain goats have traditional area-use patterns and do not increase their home range when density increases. We could have measured population density as the total number of goats in the population, the number of females two years of age and older, or simply by calendar year, as the population generally increased during the study.

Of these three measures of population density, the number of adult females is probably the most reliable because it represents the number of potential competitors for forage for a young goat of either sex. Adult males tend to use different areas from those used by nursery groups (chapter 5), and therefore their number may have a limited effect on the forage available for young goats. All three measures of density were correlated with each other and yearly changes in early-June fecal crude protein were density-independent (see box table). The number of adult females has been used to assess variability in population density in long-term studies of bighorn sheep and red deer (Clutton-Brock et al. 1997a; Festa-Bianchet et al. 1995).

We therefore used the number of females aged two years and older as our measure of population density. We compared the mass of kids, and the mass, hind foot length, and horn length of yearlings and two year olds, to the number of females two years of age and older. We could not test for the potential effects of delayed density-dependence because the strong correlation between year of study and population size meant that population size one year was strongly correlated with population density in the previous two years. Although with a longer time series the potential for delayed density-dependence could be investigated by first detrending the data, fourteen years of data are insufficient for a reliable analysis. It should be noted that the strong correlation between population size and year of study also limits our ability to separate the effects of other temporal trends (such as those attributable to climate change) and population density.

(continued on next page)

---

tion almost doubled in density, providing substantial temporal variability to compare with body and horn growth. Among ungulates, juveniles are much more susceptible than adults to density-dependent resource limitation (Fowler 1987; Gaillard et al. 1998, 2000a, 2001), therefore density-dependence in body mass should be most evident among young goats.

BOX 6.2
Continued

Correlations among different measures of population density, year, and early-June fecal crude protein content for Caw Ridge mountain goats, 1989–2003

	Year	Number of females	Crude protein
Total number of goats	0.94*	0.80*	0.09
Number of females	0.87*		−0.13
Year			−0.05

Correlations marked with an asterisk are significant at $P < 0.001$ and have a sample size of fourteen. Correlations with fecal crude protein are nonsignificant ($P > 0.7$) and sample size is twelve.

Lower body mass and horn growth were reported for well-established compared to colonizing populations of mountain goats introduced in the Olympic National Park (Stevens 1983).

Kid mass on July 15 was weakly but positively correlated with population density (table 6.1). Therefore the range of population density during our study did not have a negative effect on kid growth.

Population density had a weak negative effect on the development of yearling and two-year-old goats, but those effects varied according to both sex and age. There was no relationship between density and mass of yearling females (table 6.1), and the weak negative effect of population density on mass of yearling males[13] became not significant when examined in a multiple regression including fecal crude protein in early June. We found similar results for horn and hind foot length adjusted to July 15: population density had no effect on these measurements for females, and a weak negative effect for males that became not significant when early-June fecal crude protein was accounted for. Overall, changes in population density had little effect on the physical development of yearling goats, and the weak negative effects shown by males were possibly a consequence of year-to-year changes in plant phenology. It is interesting to note that the average masses of yearling males and females were not correlated over the years of the study,[14] suggesting that differ-

TABLE 6.1

**Relationships between the Number of Females and the Mass of
Different Sex/Age Classes of Young Goats**

Sex/age class	N	Slope	r^2	P
Kids	127	0.006	0.028	0.030
Male yearlings	72	−0.025	0.137	0.001
Female yearlings	65		0.007	0.51
2-year-old males	52	−0.041	0.195	0.001
2-year-old females	53	−0.29	0.263	0.000

Adjusted to July 15, 1989–2003. The number of females includes all females two years of age and older in the Caw Ridge population in June.

TABLE 6.2

**Regression Statistics Comparing Hind Foot Length (cm) and Horn
Length (mm) with the Number of Adult Females in June for Two-Year-
Old Mountain Goat Males and Females at Caw Ridge, 1989–2003**

Variable	Sex	b (slope)	r^2	N	P
Hind foot length	Male	−1.24	0.332	48	0.000
	Female	−1.19	0.348	47	0.000
Horn length	Male	−1.08	0.176	54	0.001
	Female	−0.73	0.105	52	0.018

Hind foot and horn lengths adjusted to July 15.

ent variables affected the physical development of yearlings of different sex.

A negative effect of population density was evident for two year olds of both sexes (table 6.1 and fig. 6.12), also when fecal crude protein in early June was accounted for.[15] Density also had negative effects on horn length and hind foot length of two year olds of both sexes (table 6.2).

We found no evidence that density affected size or growth of goats older than two years. For both sexes, mass, horn length, horn base circumference, hind foot length, and length of the third or fourth horn annuli were not correlated with population density in the year of birth. Body mass in midsummer and horn growth were also not correlated with density in that year. That result may be due to compensatory

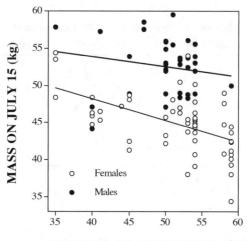

Figure 6.12. Mass of two-year-old mountain goats adjusted to July 15, according to the number of females aged two years and older in the Caw Ridge population in June.

growth, or to the fact that most goats three years of age and older monitored during our study were born before 1999, therefore in years of relatively low population density. If the population will continue to increase, we may eventually see more important consequences on body and horn growth, because the relationship between density and physical development may not be linear. So far, however, the effects of population density on the growth of mountain goats at Caw Ridge appear minimal.

Population density had a weak negative effect on mass gain by young goats. Most of the variance in body mass was not explained by density, and density had no apparent effect on the asymptotic mass, size, or horn length of goats of either sex. Therefore, the slow rate of mass gain of mountain goats at Caw Ridge does not appear to be a consequence of high population density. Instead, we suggest it is an adaptation to the very late spring and the long season with dormant forage (chapter 4). Adult mass of both sexes at Caw Ridge was equal to or greater than reported for other native mountain goat populations (Côté and Festa-Bianchet 2003). Most populations likely experience environmental conditions comparable to those on Caw Ridge or possibly even harsher; therefore, the species may be adapted to a long period of growth before reaching asymptotic mass.

Life-History Implications

Seasonal and age-specific body and horn growth of mountain goats have important implications for their life-history strategies. Those implications will be revisited in later chapters where life-history data will be presented, but it is useful to briefly summarize them here.

Mountain goats grew horns rapidly, but accumulated mass and grew in size at a slower pace than bighorn sheep in a similar environment. Among adults, sexual dimorphism in mass was substantial, but dimorphism in horn size was limited to base circumference. Therefore, assuming that sexual dimorphism is due to the selective pressure on males to be larger than potential competitors, it is likely that body mass plays a greater role than horn size (or at least horn length) in the reproductive success of males. The presence of compensatory growth in horn length of both sexes suggests that achieving the maximum possible absolute horn length may not increase fitness. Instead, it appears that selection favors a narrow range of absolute horn length.

For females, the slow growth rate over the first three years of life suggests that primiparity may occur at an advanced age, as three-year-old females are approximately 22% lighter than full-grown females. Because skeletal growth (measured by hind foot length) appears to be completed by four years of age, while mass accumulation continues until six or seven years of age (fig. 6.3), females aged six to nine years are probably in better condition than those aged four to five years. Mass loss by females older than about ten years suggests a loss of condition. These age-specific characteristics should affect the reproductive strategies of females. If old females are in poorer body condition than younger adults, they should have higher mortality and lower reproductive success.

Compared to other ungulates (chapter 8), mountain goat kids are small at weaning and possibly also at birth, suggesting either a limited effect of variability in maternal care on adult size and lifetime reproductive success, or selection for extended maternal care beyond the first year of life. The very weak sexual dimorphism at weaning also argues against the expectation of sex-differential maternal care, despite substantial adult sexual dimorphism and the likely greater role played by body size in affecting the reproductive success of males than of females.

Management Implications

The sex- and age-specific pattern of horn growth in mountain goats has major implications for their management. Part of the motivation for hunting mountain goats derives from their perceived value as trophies. Many hunters wish to shoot a goat with large horns, and a scheme exists for scoring mountain goat horns according to their size, combining various measurements of length and thickness (Stelfox 1993). Most jurisdictions, therefore, manage mountain goats as a trophy species. We know of no evidence, however, of any correlation between the size of the horns of a goat and the difficulty of the hunt. Two characteristics of mountain goat horns make their management for sport hunting difficult: horn sexual dimorphism is limited and most horn growth takes place over the first three years of life. As a consequence, untrained hunters cannot reliably distinguish males and females (Smith 1988a), and even fewer can estimate their age: from 200 m away, a four year old and a ten year old look about the same unless they stand next to each other. In nursery herds, two-year-old males are difficult to distinguish from adult females.

Most jurisdictions that allow hunting limit the total number of goats that can be taken within a management unit, then specify a minimum horn length for harvestable goats, usually in relation to ear length. A minimum horn length can prevent hunting mortality of kids and yearlings (which, ironically, are the age classes that could sustain the greatest harvest, see chapter 12), but it cannot direct the harvest to males rather than to females. For many other ungulates, the sexes are easy to distinguish in the field, and managers can restrict harvest to one sex or set sex-specific quotas. For males of wild sheep and true goats, managers may also direct the harvest toward a specific age class. These options are not available for management of mountain goats.

With some training, most hunters can develop reasonable skills in distinguishing males and females. Hunter education can increase the proportion of males in the harvest. In British Columbia 34 to 37% of goats harvested between 1980 and 1984 were females (Hebert and Smith 1986), and 50% of goats shot between 1970 and 1985 in Washington were females (Johnson 1986b), but in the Yukon, where most goats are shot by nonresidents accompanied by a guide (who have some knowledge of field identification of male and female goats), females make up about 27% of the harvest (chapter 12), suggesting that identification efforts can substantially decrease female harvests. A hunter that simply shot the longest-horned goat from a large group would almost certainly kill an adult female.

Summary

- Mountain goats of all sexes and age classes gain mass in summer and lose mass in winter. With the exception of the first winter of life, horn growth stops during winter, forming distinct annuli that give a reliable approximation of individual age.
- Mountain goat kids are small relative to the size of adult females, and growth in body size is slow. Substantial growth takes place every summer for the first four years of life, and asymptotic mass is not reached until at least six years of age. Horn growth is mostly completed by three years.
- Sexual dimorphism in body mass develops gradually and almost entirely postweaning. Adult males are about 50% heavier than adult females, but yearling males are only about 10% heavier than yearling females. Sexual dimorphism in horn length is substantial among yearlings, but over the following three years females gradually catch up to the horn length of males. Males grow thicker horns than females as yearlings and retain thicker horns through life. Dimorphism in body mass is much greater than in horn size, suggesting that male reproductive success may be influenced by body mass more than by horn size.
- There was little evidence that body and horn growth were affected by changes in population density, although both were affected by yearly changes in forage quality, particularly the timing of vegetation green-up in the spring.
- Because of the low dimorphism in horn size and shape, mountain goats present a challenge to wildlife managers and hunters that may wish to harvest adult males.

Statistical Notes

1. Test for sex difference in mass or rate of mass gain of kids over summer (capture date: $F_{1,150} = 664.6$, $P < 0.001$; sex: $F_{1,150} = 0.12$, $P = 0.73$; sex*date interaction: $F_{1,150} = 1.45$, $P = 0.23$). Capture date explained 82% of the variability in mass over all the years of the study.

2. ANOVA of the effects of sex and early June fecal crude protein (FCP) on kid mass adjusted to July 15: sex $F_{1,93} = 5.401$, $P = 0.022$; FCP: $F_{1,93} = 14.43$, $P = 0.000$.

3. Comparing mass gain of male and female yearlings over the summer: interaction sex*date in ANOVA, $F_{1,181} = 5.06$, $P = 0.026$.

4. Regression of yearling mass adjusted to July 15 on average FCP in early June of the same year: males, $F_{1,51}$ = 1.16, P = 0.28; females, $F_{1,43}$ = 0.001, P = 0.98.

5. From repeated mass measurements of the same individuals, comparing summer mass gain rate of two-year-old males and females (mean ± SD; males: 308 ± 66 g/day, n = 20; females: 237 ± 96 g/day, n = 14; t_{32} = 2.56, P = 0.016).

6. Correlation of average mass at six to eight years of age and longevity for females that died of natural death or were alive but aged nine years and older in 2003: r = 0.14, n = 34, P = 0.41.

7. Increase in hind foot length during summer for forty-seven two-year-old females (0.2 mm per day, r^2 = 0.088, P = 0.04) and forty-nine males (0.15 mm per day, r^2 = 0.054, P = 0.10).

8. Yearling horn length on July 15 compared to fecal crude protein in early June (males: 93 mm + 2.4 (FCP), n = 53, P = 0.006, r^2 = 0.14; females: 82 mm + 1.9(FCP), n = 45, P = 0.04, r^2 = 0.09).

9. Correlation of yearling horn length on July 15 and body mass (males: r^2 = 0.39, n = 71; females: r^2 = 0.28, n = 63, both P < 0.001).

10. ANOVA comparing yearling horn length with June fecal crude protein, including the effects of sex and mass: $F_{1,94}$ = 25.07, P < 0.001.

11. Regression of horn length and date for goats aged three years and older (males: r^2 = 0.033, P = 0.41, n = 22; females: r^2 = 0.0, p = 0.99, n = 15).

12. Horn annuli by sex: males have longer first annuli (t_{193} = 11.25, P < 0.0001), females grow longer annuli than males in the following two years (t_{102} = 5.38, P < 0.001 and t_{69} = 5.14, P < 0.001; fig. 6.8). Beyond four years of age, females appeared to grow longer annuli than males, the difference was marginally significant at ages four and five (P = 0.04 and P = 0.047) and not significant at ages six and seven (P's > 0.6).

13. ANOVA including the effects of sex, density and their interaction on yearling mass on July 15. The sex*density interaction was significant ($F_{1,135}$ = 4.42, P = 0.037), therefore males and females were considered separately. There was no effect of density on mass of yearling females (P = 0.51). Males showed a weak negative effect of density on mass (r^2 = 0.14, b = –0.025, n = 72, P = 0.001). For horn and hind foot length adjusted to July 15, density had no effect for females, and a weak negative effect for males (horn length: r^2 = 0.07, b = –0.701, n = 73, P = 0.02; hind foot length, r^2 = 0.06, b = –0.53, n = 66, P = 0.046).

14. Correlation between the yearly average masses of yearling males and females, adjusted to July 15, over 14 years: r = 0.07, P > 0.1.

15. For two year olds, no significant sex*density interaction on mass ($P =$ 0.39). We examined the relationships between density and mass separately according to sex, because of the substantial sexual difference in body mass among two year olds. The negative effect of density on the midsummer mass of two year olds of both sexes persisted when examined in an ANOVA with fecal crude protein in early June (density: $F_{1,78} =$ 15.01, $P = 0.000$; fecal protein: $F_{1,78} = 8.08$, $P = 0.006$).

CHAPTER 7

Individual Variability in Yearly and Lifetime Reproductive Success of Females

A major contribution of behavioral studies to both fundamental ecological theory and to conservation biology is the emphasis on individual variability, and on identifying the sources of that variability (Gosling 2002). Conservation plans that simply assume equal reproductive success among all individuals, or random variation in reproductive success, are unlikely to succeed (Coulson et al. 2001; Gaillard et al. 2003) because in real populations there are many individual differences, some of which can be associated with differences in sex–age classes. It is therefore important to identify the extent and the sources of variability in individual reproductive success.

Although numerous studies of ungulates have examined the correlates of seasonal or short-term reproduction, very few have monitored marked individuals for a sufficiently long period to establish variability in lifetime reproductive success. In this chapter we examine how the reproductive success of mountain goat females is affected by age, mass, and parturition date. We then measure individual variability in lifetime reproductive success and compare it to differences in female longevity. In the following chapter we build upon the results presented here to examine costs of reproduction and female reproductive strategy.

Female Age and Reproductive Success

Much recent work on the population dynamics of large herbivores has emphasized the effects of a population's age structure on recruitment and survival. Because reproductive success in female ungulates is strongly age-dependent, age structure can have a major effect on recruitment rate (Bérubé et al. 1999; Coulson et al. 2001; Gaillard et al. 2000a, 2001). Consequently, we begin our exploration of the causes of variability in female reproductive success with an examination of the effects of age.

After a few years of monitoring known-age mountain goats, it became apparent that females did not begin to reproduce until a surprisingly late age (Festa-Bianchet et al. 1994). Based on the body mass of adult females, we expected that primiparity would occur at two or three years, as reported for similar-sized ungulates such as bighorn sheep (Jorgenson et al. 1993a), red deer (Langvatn et al. 1996), and fallow deer (Birgersson 1998a). Indeed, in introduced mountain goat populations, most three year olds and some two year olds produce kids (Bailey 1991; Houston and Stevens 1988). Instead, mountain goats at Caw Ridge never gave birth at two years and very exceptionally produced kids at three years of age. Of forty-five females born before 2000 whose age of primiparity was known, 45% did not reproduce until five years of age or older (fig. 7.1). Young females spent several years without contributing to recruitment, and 27% of fifty-five females that survived to two years of age died or emigrated before reproducing on Caw Ridge. This analysis excluded the twenty-five females that were captured and drugged when aged three or four years, because chemical immobilization of young females had a negative effect on their probability of reproducing the following year (Côté et al. 1998a). Nevertheless, over a quarter of females that reached "adult" age

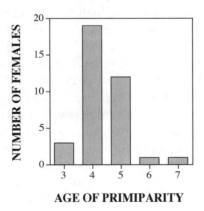

AGE OF PRIMIPARITY

Figure 7.1. Age of primiparity of mountain goat females born at Caw Ridge from 1988 to 1999. Females that were chemically immobilized when aged three or four years were excluded.

(most studies of ungulates would consider two-year-old females as adults) did not contribute any recruitment to the population.

The late age of primiparity in mountain goats on Caw Ridge appears surprising considering the asymptotic mass of adult females, but not when considering the slow rate of growth of young mountain goats presented in chapter 6 and the age-specific mass of young females. A comparison with bighorn sheep reveals that the key factor is the age at which females achieve a substantial proportion of asymptotic mass. When they first conceive at eighteen or thirty months of age, most bighorn ewes weigh 55 to 65 kg, or 77 to 92% of the average mid-September mass of adult ewes (Gallant et al. 2001). In contrast, assuming a linear mass gain until September, most female mountain goats aged eighteen or thirty months only weigh 42 to 55 kg, or 54 to 70% of the mass of adult females at the same time of the year (fig. 6.4). It is therefore not surprising that they first conceive at 3.5 or 4.5 years of age, when they reach about the same proportion of adult mass that bighorn ewes reach one year earlier. Therefore, primiparous bighorn and mountain goat females have about the same body mass relative to fully grown adults, but goats reach that mass one or two years later than sheep. In ungulates generally, primiparity does not occur at a set age, but at the age when females reach about 80% of their asymptotic mass (Sadleir 1987). Both bighorn sheep and mountain goats appear to support this general rule, although we currently do not have enough data to test whether heavy young mountain goat females reach primiparity earlier than light ones.

We knew from field observations whether or not a female's kid survived to September, but because many kids were unmarked we could not determine their survival to one year. In mid-May, just before the birth of new kids, most but not all surviving yearlings were still associated with their mothers. In addition, in a few years fieldwork did not begin until late May. When the postweaning survival of an individual kid was unknown, we assigned to each mother a proportion of the unmarked kids with unknown mothers that had survived the winter. In years when we knew the sex of all kids, we assigned different "kid fractions" to mothers known to have had a male or a female kid, based on sex-specific overwinter survival. For example, if in early June there were eight mothers whose kid had survived to September the previous year, and six unmarked yearlings with unknown mothers, we would assign 0.75 kids to each of those eight mothers. This procedure tended to smooth out age-specific and individual differences in reproductive success, but provided a more accurate estimate of kid survival to one year than if we simply ignored yearlings with unknown mothers. We assigned a "fraction of a kid" to sixty-five female-years, or 15% of the total.

The proportion of females giving birth increased from three to six years, peaked at about 81% from eight to twelve years of age, then declined slightly for females thirteen years of age and older (fig. 7.2). Mother age, however, did not seem to affect kid survival to either weaning or to one year of age (fig. 7.2 panels B and D; Côté and Festa-Bianchet 2001d). Differences in age-specific kid recruitment, therefore, were caused mostly by changes in fecundity, not kid survival. When we

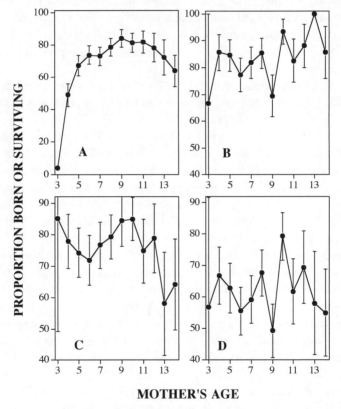

MOTHER'S AGE

Figure 7.2. Age-specific reproduction by mountain goat females at Caw Ridge, 1989 to 2003. Points represent means (± SE) for each female aged from three to thirteen years. The last point shows the mean for females aged fourteen years and older: (A) kid production, (B) kid survival to weaning, (C) kid survival from weaning to one year of age, (D) kid survival from birth to one year of age. Note changes in scale.

examined yearling recruitment on a per capita basis, we found that, despite an unexpected drop in reproductive success among nine year olds, females aged eight to twelve years were more likely to produce a yearling (mean of 0.51 yearlings/year) than females that were either younger (0.41 yearlings/year, for females aged four to seven years) or older (0.37 yearlings/year). These differences underline the important role of mature females in population dynamics. Females aged eight to twelve years were about 30% more likely than other females to recruit a yearling in the population. This is an important consideration for mountain goat management. Some wildlife managers tend to assume that young females are more productive than mature females. For mountain goats, that clearly is not the case, possibly because of the long period of body growth (chapter 6). Fully grown but presenescent females, aged eight to twelve years, are likely in the best body condition, have more breeding experience, and enjoy high reproductive success.

Reproductive senescence appeared to set in at about thirteen years (fig. 7.2). Kid production by females aged thirteen years and older was 67% (n = 43 female-years) compared to 81% (n = 205 female-years) for those aged eight to twelve years. That comparison, however, assumes that all adult females have similar probabilities of surviving to thirteen years of age. Although we did not find any evidence that body mass as a young adult (a measure of phenotypic quality) was related to longevity, the possibility remains that females surviving to old age were of superior phenotype compared to those that died young (Gaillard et al. 2000b), because phenotypic quality may be related to variables other than body mass as a young adult. If survival to old age increased with phenotypic quality, the comparison of reproductive success by prime-aged and senescent females may have underestimated the negative effects of senescence (Partridge and Mangel 1999). The figures reported above on age-specific kid production are of interest from a population dynamics viewpoint because they can relate a population's age structure to productivity but may not provide a reliable measure of how female reproductive potential changes with age.

A more direct assessment of reproductive senescence would be a comparison of kid production by the same female at different ages. For seventeen females, we monitored kid production between eight and twelve years of age (mean 80%) and at thirteen years of age and older (mean 70%). Although not significant,[1] the age-related decline was very similar to that shown by the population as a whole. Lack of significance was probably a consequence of small sample size. Kid production between eight and twelve years for these long-lived females was similar to that for the overall population, suggesting that females surviving to thirteen years and older were not phenotypically superior to females that survived to at least eight years of age.

Reproductive senescence is unlikely to have a strong impact on population dynamics of mountain goats, for two reasons. First, the decrease in kid production for females aged thirteen years or older is not very strong. Most old females produce kids and those kids enjoy the same survival as those born to younger mothers. The second and more important reason is that few females survive to the age where senescence manifests itself. From 1999 to 2003, when the age of almost all females was known with certainty, those aged thirteen and older made up only 16% of females aged more than three years. Our data strongly suggest that there is no reproductive senescence in female mountain goats up to at least twelve years of age and provide no support for any management strategy aimed at improving productivity by removing old females.

When kid production and survival were combined (fig. 7.3), it became evident that females aged three years and younger contributed almost no recruitment and females aged four years had lower reproductive success than older females. Therefore, the population included many young females that contributed very little to recruitment, particularly if one includes two year olds, none of which reproduced. From 1989 to 2003,

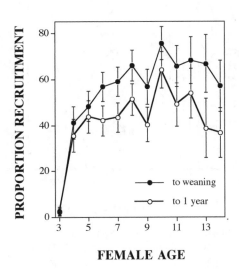

FEMALE AGE

Figure 7.3. Recruitment (± SE) of weaned and one-year-old kids by female mountain goats at Caw Ridge, 1989 to 2003. The point for age fourteen includes all females aged fourteen years and older.

females aged two to four years made up a third of the total number of marked females aged two and older. Females aged three or four years contributed 24% of those aged three years and older. Although all females aged three years and older would be classified as adults (Smith 1988a), as much as one quarter would contribute little recruitment because of their young age.

Kid production by young females was strongly affected by maternal dominance status. Between 1994 and 1997, eight of the eleven kids produced by females aged three to six years were born to mothers in the top 35% of the dominance hierarchy for females of that age range. For older females, the effects of dominance on kid production were also positive, but much weaker (Côté and Festa-Bianchet 2001d) as most females produced a kid in most years.

To examine the effects of maternal age on kid mass we grouped mothers into three age classes: young mothers aged three to five years that have not completed body growth (chapter 6), prime-aged mothers of six to twelve years, and mothers aged thirteen years and older that may show reproductive senescence. Maternal age had a weak effect on kid mass, with kids born to young and old mothers being 8 to 9% lighter in mid-

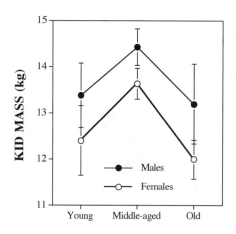

MATERNAL AGE

Figure 7.4. Mass adjusted to July 15 (kg ± SE) of mountain goat kids born to mothers of different ages at Caw Ridge, 1989 to 2003. Young mothers were aged three to five years, middle-aged mothers were six to twelve years old, and old mothers were aged thirteen years and older. Sample size was 15 males and 16 females for young mothers, 38 males and 42 females for middle-aged mothers, and 4 males and 2 females for old mothers.

July than those born to prime-aged mothers (fig. 7.4), but the effect was not significant, possibly because of strong year effects on kid mass.[2] This analysis also suggested that when maternal age was taken into account, male kids were about 6% heavier in midsummer than female kids (fig. 7.4). Because the sample included only six kids born to old mothers, it is likely that the main source of the difference was in the contrast between kids of young and of prime-aged mothers. Although a larger sample of older mothers is required to better assess the effect of maternal age on kid mass, the apparently minor effect that we recorded agrees with the observation that maternal age had little effect on a kid's probability of survival. Instead, maternal age affected reproductive success almost exclusively by affecting the probability of kid production. Because we were rarely able to document neonatal mortality, however, we cannot exclude that maternal age may affect kid survival in the few hours or days following birth.

Influence of Mass and Body Condition on Reproductive Success

In many ungulates, female body mass is positively correlated with reproductive success although the strength of that correlation is variable (Adams and Dale 1998; Cameron et al. 1993; Festa-Bianchet et al. 1998). To determine whether or not maternal mass may affect reproductive success of mountain goats, we adjusted mass for capture date and accounted for differences in age, two variables that had a considerable effect on body mass (chapter 6).

We compared the mean mass of females adjusted to July 15 at six to ten years of age to their mean yearly reproductive success during those years. Female mass does not vary much between six and ten years (fig. 6.4) and the mean mass of females aged six to ten years did not vary among years.[3] We thus considered thirty-seven females that were weighed at least once when aged six to ten years and for which we had at least two years of reproductive data. The relationship between mass in midsummer and reproductive success was generally positive, but it was weak and not significant. Mass was not correlated with annual kid production, nor with yearly weaning success.[4] These results are somewhat inconclusive because of the small sample size but do not suggest a strong role of female mass on reproductive success.

A female's ability to reproduce may be affected more by her body condition than by her absolute mass, because the ability to provide maternal care should partly depend on the availability of energy reserves in the form of fat deposits (Festa-Bianchet 1998; Gerhart et al. 1997; Testa and Adams 1998). We estimated body condition at capture for twenty-six females aged six to ten years, by dividing mass (adjusted to mid-July) by foot length. This measure of condition was weakly but significantly correlated with weaning success, but not with the yearly probability of producing a kid.[5] Despite the small sample, these results suggest that energy reserves may play a role in female reproductive success.

Overall, there was no strong evidence suggesting that heavy adult female were more likely than light females to either produce or wean kids. Because most relationships examined had positive correlation coefficients, it seems reasonable to suspect that body mass had a slight positive effect on reproductive success. That result is somewhat surprising given the wide variability in individual mass and body size among the females included in this analysis. Average mid-July mass of adult females was 72.5 kg (range 63.1–81.6). Hind foot length averaged 345 mm (range 317–378). The largest female was almost 30% heavier and 19% taller than the smallest female. That variability should have been adequate to detect a moderate effect. It appears that body mass explained at best about 10% of the variability in the reproductive success of adult female mountain goats. Given the small size of kids and their slow rate of growth, a possible explanation for this result is that the reproductive effort of mountain goat females is limited, so that reproduction is not much more difficult for small than for large females, unlike what has been suggested for bighorn sheep (Festa-Bianchet et al. 1998; Réale and Festa-Bianchet 2000). It is also possible that maternal size effects would be more evident in years with greater resource scarcity than during our study. We will return to this theme in the next chapter.

Birthdate and Kid Survival

For many temperate ungulates, parturition dates in the same population are highly synchronized, including bighorn sheep (Festa-Bianchet 1988c), red deer (Guinness et al. 1978), moose (Bowyer et al. 1998; Keech et al. 2000), and roe deer (Gaillard et al. 1993b). Often, over 80% of births take place during a two-week period in spring. Two hypotheses have been proposed to explain these short parturition seasons, invoking either predation or plant phenology as the selective pressure. The "predation" hypothesis suggests that highly synchronized births are an adaptation to reduce predation on neonates. Newborns are highly vulnerable to predation, but if most young are born during a short time, their survival may improve because predators will be swamped by the number of newborns

(Dauphiné and McClure 1974; Estes 1976; Gregg et al. 2001; Linnell et al. 1995). If births were spread over time, more newborns could be killed and eaten by predators than if births were synchronous, because vulnerability to predation declines as the young age. Neonates born before or after the main birth pulse would face a high risk of predation because they would not be protected by predator dilution. This hypothesis is particularly relevant if predators are more effective at killing newborns than older juveniles. For mountain goat kids, this category could include both eagles and grizzly bears. Eagles are likely limited by the weight of the prey that they can lift and grizzly bears are often very efficient predators of ungulate neonates (Larsen et al. 1989). Therefore, it is reasonable to suspect that the risk of predation by eagles and bears for newborn goats decreases rapidly as they age and gain both mass and agility.

The "phenology" hypothesis relates birth synchrony to the strong seasonality of plant phenology in temperate environments. Seasonal changes in plant phenology affect the quality and availability of forage and may constrain the timing of parturition. Mothers that gave birth just before the onset of vegetation growth would feed at the seasonal peak in forage availability and quality at the time when they must meet the high energetic demands of lactation. In addition, their offspring could feed on high-quality forage as they gradually switch from milk to solid foods (Bunnell 1980, 1982; Rachlow and Bowyer 1991; Thompson and Turner 1982).

The predation and phenology hypotheses are difficult to separate because they are not mutually exclusive. Even if predator swamping is important and favors synchronous births, other selective pressures should still be expected to result in the peak in parturitions to be timed to vegetation phenology. The two hypotheses, however, make different predictions about the effects of birthdate on juvenile survival at different ages. The predation hypothesis predicts that late-born juveniles will suffer high neonatal mortality but does not predict any difference in survival after the first few days of life. On the other hand, the phenology hypothesis makes no predictions about neonatal survival but predicts that late-born juveniles will suffer heavy mortality in the fall or during winter, because they will not be able to accumulate sufficient body mass and fat reserves during the vegetation growing season. Differences in neonatal mortality relative to birthdate would therefore support the predation hypothesis, while birthdate-related differences in juvenile survival in autumn or winter would support the phenology hypothesis.

The predation hypothesis would be supported by evidence that females synchronize their breeding cycle with that of other females within their group. Female red deer and sable antelope may synchronize estrus with other females in their population (Iason and Guinness 1985; Thompson 1995) and late-breeding female bison in good condition appear to shorten their gestation in order to give birth near the peak time of parturitions (Berger 1992). In support of the phenology hypothesis, late-born juveniles of several ungulate species in temperate environments have low overwinter survival (Birgersson and Ekvall 1997; Fairbanks 1993; Festa-Bianchet 1988c; Guinness et al. 1978; Keech et al. 2000; Smith and Anderson 1998). The interpretation of patterns of survival according to birthdate, however, is complicated by the effects of maternal age and condition: young mothers and mothers in poor condition tend to give birth later than prime-aged females or females in good condition, likely because good body condition leads to earlier ovulation (Birgersson and Ekvall 1997; Cameron et al. 1993; Fairbanks 1993; Festa-Bianchet 1988b). Therefore, lower survival of late-born juveniles may be due partly to poor maternal quality rather than simply to a mismatch between birthdate and phenology.

From 1993 to 1999 we intensified efforts to determine kid birthdate. We searched the study area intensively from mid-May to early June, trying to locate each adult female every day. In the following analyses, we included only females for which we were confident that our estimated parturition date was correct within five days. The estimated date of the first parturition each year ranged from May 14 to 22 (table 7.1). There

TABLE 7.1
Descriptive Statistics of the Parturition Season of Mountain Goats, 1993–1999

Year	Birthdate				Length of birth Season (days)	80% of births[a]		
	Mean	Median	First kid	Last kid		Date	# Days	N^b
1993	01/06	27/05	19/05	03/07	46	07/06	19	14
1994	28/05	24/05	21/05	19/06	30	02/06	11	23
1995	29/05	27/05	22/05	26/06	36	03/06	11	28
1996	28/05	25/05	20/05	24/06	36	02/06	13	21
1997	26/05	24/05	20/05	16/06	28	27/05	7	23
1998	20/05	20/05	14/05	08/06	26	23/05	9	35
1999	24/05	22/05	17/05	12/06	27	26/05	9	26
All years	26/05	24/05	14/05	03/07	33	30/05	16	170

[a]Number of kids with known birthdate each year
[b]From Caw Ridge, Alberta. (Updated from Côté and Festa-Bianchet 2001a).

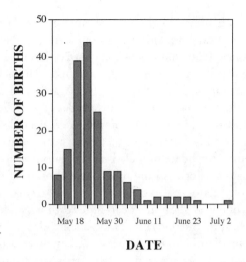

DATE

Figure 7.5. Birthdates of 170 mountain goat kids in 1993–1999 at Caw Ridge, Alberta. Histograms indicate the number of kids born every three days beginning on May 14.

was no difference in birthdate according to kid sex: median birthdate was May 24 for seventy-five males and May 25 for sixty-eight females.[6] The parturition season (the time from first to last birth) lasted thirty-one days on average (table 7.1) but the distribution of births included an early peak followed by a few very late births (fig. 7.5).

Because of the skewed distribution of births, the median birthdate provides a better indication of year-to-year differences than the mean. The overall median birthdate was May 24 and ranged from May 20 to 27. Parturitions were highly synchronized: 80% of all kids were born within seventeen days of the first birth. Only 16% of kids were born on or later than June 1 (table 7.1, fig. 7.5). We have already examined the effects of maternal age, social rank, breeding experience (primiparous versus multiparous) and population density on kid birthdate. None of those variables had significant effects, even when analyses were carried out separately by kid sex (Côté and Festa-Bianchet 2001a). There was no evidence that mothers of late-born kids were young or in poor condition. Possibly, some females failed to conceive on their first estrus and cycled again, leading to late births (Bunnell 1980).

We also explored the effects of birthdate on kid mass, for the subsample of kids with known birthdate and at least one mass measurement. After adjusting for differences in capture date, a multiple regression revealed that birthdate, fecal crude protein, and the mother's previous

breeding experience affected kid mass, explaining 35% of the variance.[7]
By mid-July, the earliest-born kids were about 4 kg (or 25%) heavier than
the latest-born ones (Côté and Festa-Bianchet 2001a).

In the Sheep River population of bighorn sheep, lambs born late suffered much higher mortality than those born early, and differences in
survival were evident for lambs born only ten days apart (Festa-Bianchet
1988c). Because of the similar environment, we expected similar results
for mountain goat kids but our expectation was confirmed only for
males.[8] The median birthdate of male kids that survived to weaning was
May 23, eight days earlier than the median for those that died. Surprisingly, however, birthdate had no effect on survival to weaning for female
kids. Survival to one year showed a near-significant negative relationship
with date of birth: surviving kids had a median birthdate of May 23, those
that died had a median birthdate of May 26, but there was no effect of
birthdate on survival to one year when the sexes were considered separately.[8]

The weak negative effect of birthdate on kid survival was evident when
kids born after June 14, three weeks after the median birthdate, were
considered. Of these ten very late-born kids, seven survived to weaning
and five to one year. In the same years, overall kid survival to weaning was
81% and survival to one year 63%. These results contrast with the dramatic effects of late birth on survival of bighorn lambs at Sheep River,
where only 5% of those born after June 10 survived to one year (Festa-
Bianchet 1988c).

Selection to avoid predation on newborns was unlikely to explain the
highly synchronous parturition of mountain goats. Predation on neonates was very rare during our study. We did not directly document any
predation on newborns, and 95% of kids known to have been born survived to two weeks of age, including thirty-seven of thirty-nine kids born
in June or later. There was no evidence that late-born kids suffered
greater neonatal mortality than kids born during the peak parturition
season. Parturient females distribute themselves widely over the ridge
(Côté and Festa-Bianchet 2003), making newborn kids a very unprof-
itable prey to search for, particularly given that there are few alternative
prey species available on alpine ridges in late May and early June. The
only predator that could efficiently search for newborns was the golden
eagle, but the high neonatal survival suggests that eagle predation was
very rare. Despite thousands of hours of observations over several years,
we saw only three serious predation attempts by an eagle on kids. We
never had the impression that golden eagles actively searched Caw Ridge

during the parturition season, although females with kids reacted to eagles as if they perceived them as a threat. Kids ran underneath their mothers when an eagle approached.

The most important benefit enjoyed by early-born kids was access to high-quality and abundant forage for a long period, as suggested by the positive effect of birthdate on kid mass in midsummer (Côté and Festa-Bianchet 2001a). That mass advantage, however, appeared to translate into greater survival to weaning for males only, and survival to one year was independent of birthdate for both sexes. Compared to bighorn sheep at Sheep River or Ram Mountain, mountain goats on Caw Ridge had access to high-quality forage for a rather long period of time (fig. 4.5), which may have allowed for some compensatory growth by late-born kids. Even in the absence of compensatory growth, late-born kids may have reached a threshold mass that allowed a good chance to survive the winter. The substantial differences in mass between early- and late-born kids that we detected in mid-July may have narrowed by early autumn. Alternatively, size at the onset of winter may not affect the survival of mountain goat kids as much as that of bighorn lambs. Finally, the trend in the data, although not significant, is in the expected direction: late-born kids do rather worse than early-born ones. Perhaps a larger sample is required to draw a firmer conclusion about the effects of birthdate on kid survival. It is likely that those effects will be stronger in unfavorable years, with harsh weather or high population density.

Lifetime Reproductive Success

Similar to other ungulates, adult mountain goat females enjoy high survival (chapter 8). It takes many years for a cohort of females to provide complete data on lifetime reproductive success. After fifteen years of monitoring, we have complete or near-complete data for only a few cohorts. Because parturition at three years of age was very rare (fig. 7.1), we assumed that we had complete lifetime reproductive success data for females monitored from when aged four years or younger. We thus documented complete lifetime reproductive success of twenty-two females born before 1992 that survived to at least four years of age. Because we found no evidence that age of death was related to individual quality, we also included four females born in 1992 to 1994 that survived to at least four years and had died by 2003. This sample included five females alive in June 2004, aged on average 14.6 years (range 13–17), close to their maximum life span. For 19 of 120 kids of these females that survived to

TABLE 7.2

Lifetime Reproductive Success for Twenty-six Mountain Goat Females

Lifetime number of kids	Mean	SD	Range
Produced	5.7	3.04	0–10
Weaned	4.6	2.70	0–9
To 1 year	3.6	2.42	0–9
Nanny longevity (years)	11.0	3.69	4–18

For females that survived to at least four years and either died naturally (n = 21) or were alive but aged at least thirteen years in 2004 (n = 5).

TABLE 7.3

Relationships (r^2, coefficients of determination) between Longevity and Three Measures of Lifetime Reproductive Success in Female Bighorn Sheep and Mountain Goats in Alberta, Canada

Number of offspring	Produced	Weaned	Surviving to one year
Bighorn ewes (SR)	0.828	0.769	0.480
Bighorn ewes (RM)	0.893	0.731	0.516
Mountain goats	0.826	0.733	0.774

Sample sizes were 26 mountain goats that died of natural causes when 4 years and older or were aged at least 13 years in 2004 at Caw Ridge, 135 bighorn ewes that died at 2 years and older at Ram Mountain (RM) and 67 bighorn ewes that died at 2 years and older at Sheep River (SR). Mountain goats that died at 2 or 3 years were excluded because no 2-year-olds and very few 3-year-olds reproduce.

weaning, we did not know whether or not they survived to one year and we assigned to their mothers a fraction of kid equal to the sex-specific proportion of unmarked kids that survived the winter in that year.

Lifetime reproductive success varied considerably among females (table 7.2). Because goats in this sample produced only one kid a year, none of the variability in reproductive success could be attributed to changes in litter size. With litter size fixed at one, lifespan should strongly affect lifetime reproductive success. We examined the relationship between longevity and the number of kids produced and weaned for the twenty-six nannies that survived to at least four years of age with complete (or near-complete) lifetime reproductive records. As expected, longevity was positively and strongly associated with lifetime reproductive success (figs. 7.6 and 7.7). The correlation between female longevity and reproduction weakened when reproductive success was measured as

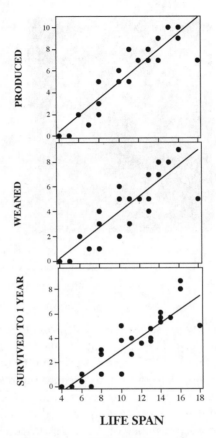

Figure 7.6. Relationship between life span and three measures of reproductive success (number of kids produced, weaned, and that survived to 1 year of age) for 26 mountain goat females on Caw Ridge that survived to at least 4 years of age and that either died of natural causes or were aged at least 13 years in June 2004.

the number of kids weaned or the number of kids surviving to one year (fig. 7.6). That decrease in the predictive power of longevity was expected, as kid survival was not affected by maternal age (fig. 7.2) and kid mortality introduced additional variability to the data (Cabana and Kramer 1991). A comparison with bighorn ewes from two populations in Alberta, however, revealed that the association between female longevity and lifetime reproductive success was stronger in mountain goats than in bighorn sheep (table 7.3). In both species and in all three populations, there was a strong relationship between life span and the number of offspring born or weaned, but life span in mountain goats explained almost

Figure 7.7. Longevity was the main determinant of lifetime reproductive success in females. Female #8 died at fifteen years of age and weaned eight kids during her lifetime. Photo by S. Côté.

twice as much variability in the number of offspring that survived to one year than in bighorn sheep. This result implies that for mountain goat females, surviving as long as possible and continuing to produce one kid each year was the major determinant of lifetime reproductive success. Factors affecting kid survival, such as varying amounts of maternal care or environmental influences, were less important than for bighorn lambs. The relationship between number of offspring produced and number surviving to one year over a female's lifetime seemed stronger in goats ($r^2 = 0.83$, $N = 25$) than in bighorns (Ram Mountain: $r^2 = 0.66$, $N = 135$; Sheep River: $r^2 = 0.60$, $N = 67$). These results are partly a function of the observation that on average kid survival was higher than lamb survival (Gaillard et al. 1998).

There are two alternative interpretations of these interspecific differences. Perhaps there is less individual variability in the amount of maternal care provided by mountain goat females than by bighorn ewes. It appears that mountain goats provide a generally lower amount of maternal care than other ungulates. Another interpretation is that during the fifteen years when we monitored mountain goats at Caw Ridge the population did not encounter the conditions that were experienced by bighorn sheep at Sheep River and Ram Mountain, which led to greater variability in lamb than in kid survival. Although predation levels likely varied from year to year at Caw Ridge, they did not seem to change as much as cougar predation in both bighorn sheep study areas (Festa-Bianchet et al. 2006; Ross et al. 1997). In addition, the Ram Mountain population displayed strong density-dependence in lamb survival (Portier et al. 1998), and the Sheep River population was affected by a pneumonia epizootic (Festa-Bianchet 1988a). Density-dependence, high cougar predation, and pneumonia epizootics had strong effects on bighorn lamb survival, but no comparable event was recorded during our mountain goat study.

Summary

- Most mountain goat females on Caw Ridge attained primiparity at four or five years of age, later than expected based on asymptotic adult mass. Because they gained mass more slowly than other ungulates, however, the mass of primiparous goats relative to asymptotic female mass was similar to that of other primiparous ungulates.
- Kid production increased with female age from four to six years, peaking at 80% at eight to twelve years. Fertility appeared to decline after twelve years of age. Kid survival to one year of age was independent of maternal age.
- Because of the late age of primiparity and increasing kid production with age, much of the recruitment of yearlings to the population was contributed by females aged eight to twelve years. In any one year, approximately a third of adult females (aged three years and older) did not produce kids.
- Maternal mass and condition had positive but very weak effects on kid survival, possibly because all mothers were able to provide a sufficient amount of maternal care.
- Over 80% of kids were born within seventeen days of the first birth. Yearly median birthdate ranged from May 20 to May 27. Late-born kids were smaller in midsummer than kids born during the parturition peak, but birthdate did not affect survival to one year. The highly synchronized parturition season of mountain goats appears to be an adaptation to a highly seasonal environment.
- Lifetime reproductive success of female mountain goats was strongly affected by longevity. Long-lived females were able to produce many kids. Female longevity had no obvious effect on kid survival.

Statistical Notes

1. Comparison of yearly kid production each year between eight and twelve years of age (mean 80%) and for an average of 2.4 years/female at thirteen years of age and older (mean 70%) for seventeen females: Wilcoxon matched-pairs sign rank test, $z = -1.44$, $P = 0.15$.

2. ANOVA assessing the effects of maternal age (three categories: mothers aged 3–5, 6–12, and 13+ years), year of birth, and sex on kid mass in mid-July: year, $F_{13,110} = 2.363$, $P = 0.008$; maternal age, $F_{2,110} = 2.055$, $P = 0.13$; kid sex, $F_{1,110} = 3.588$, $P = 0.061$.

3. ANOVAs comparing female mass between six and ten years of age: $F_{4,75} = 0.43$, $P = 0.78$ and the mean mass of females aged six to ten years among years (including only years when at least four females in this age range were weighed, $F_{8,68} = 1.51$, $P = 0.17$, yearly range 68.7 to 76.7 kg).

4. Correlations of mass in midsummer with annual kid production ($r_s = 0.28$, $P = 0.09$) and with yearly weaning success ($r_s = 0.16$, $P = 0.34$).

5. Correlations of body condition (mass divided by foot length) with weaning success at six to ten years ($r_s = 0.41$, $P = 0.039$, $N = 26$) and with the yearly probability of producing a kid at six to ten years ($r_s = 0.30$, $P = 0.14$, $N = 26$).

6. Z-transformation of Mann-Whitney U test for birthdate according to kid sex: median May 24 for 75 males, May 25 for 68 females ($Z = -0.818$, $P = 0.41$).

7. Multiple regression of the residuals of a linear regression of kid mass on capture date with maternal age, age-specific maternal social rank, previous breeding experience, kid birthdate, fecal crude protein in June, and population density. Only birthdate ($B = -0.41$, $P = 0.008$), fecal crude protein ($B = 0.43$, $P = 0.004$), and previous breeding experience ($B = 0.27$, $P = 0.06$) affected kid mass, explaining about 35% of the variance (final model: $F_{3,33} = 5.99$, $P = 0.002$).

8. Logistic regression of survival to weaning compared to birthdate (May 15 = day 0) for male kids $b = -0.068$, $P = 0.036$, $N = 73$ and for female kids $b = -0.01$, $P = 0.8$, $N = 67$. Logistic regression of survival to one year compared to birthdate for kids of both sexes $b = -0.035$, $P = 0.07$, $N = 146$, for males: $b = -0.041$, $P = 0.11$, $N = 64$, and for females: $b = -0.026$, $P = 0.44$, $N = 58$.

CHAPTER 8

Female Reproductive Strategy

The idea that natural selection favors the "fittest" individuals is deceptively simple. The measurement of fitness is the subject of much theoretical debate (Arnold and Wade 1984; Steen 1983; Wade and Kalisz 1990). An adequate estimation of fitness has proven very difficult for free-ranging populations of wild animals (Benton and Grant 1999; Brodie and Janzen 1996; Kruuk et al. 2000; McGraw and Caswell 1996). At first glance, it seems obvious that beneficial genes should spread within a population, while inferior genes should be weeded out. For iteroparous organisms that distribute reproductive effort over several reproductive episodes, it may also seem relatively clear that selection should favor an optimal allocation of resources to growth, survival, and reproduction so as to maximize fitness. When generations overlap, however, the distribution of reproductive effort over time and the growth rate of the population affect the measurement of fitness, which is further complicated by covariation among life-history traits (Cam et al. 2002) and density-dependent changes in the fitness costs of reproduction (Clutton-Brock et al. 1997b; Coltman et al. 1999b). A recent innovation (Coulson et al. 2006) provides a practical way to estimate individual fitness by considering the relative impact of individual performances on population dynamics on a yearly basis.

The concept of "reproductive strategy" implies that individuals will not simply attempt to maximize their instantaneous reproductive output, but should rather allocate resources to reproduction taking into account the possible effects of current reproductive effort on future reproductive potential. In some cases, the best strategy may be not to reproduce at all, if by so doing an individual may substantially increase its average chances of high reproductive success at a future time. Mountain goat females that begin to reproduce before completing body growth are likely to face environmentally induced changes in the fitness costs of reproduction and can expect strong age-related changes in survival probability and reproductive potential. Their reproductive strategy therefore should be affected by their age, body condition, and resource availability. When resource availability changes unpredictably, it is unlikely that natural selection will favor a fixed optimal reproductive strategy (Clutton-Brock et al. 1996). Instead, individuals should be selected to change their allocation of resources to reproduction and to other vital functions according to their physical state at each reproductive event (McNamara and Houston 1996), showing substantial phenotypic plasticity. In a highly seasonal environment, females face a large variation in resource availability over the course of a single reproductive event: a mountain goat female in average body condition that conceives in late November cannot predict the harshness of the upcoming winter, the timing of snowmelt in the following spring when she will give birth, or the vegetation quality during the summer when she will produce milk and care for her kid.

Here we address three questions concerning the evolution of reproductive strategy in female goats. First, we examine why not all females begin reproducing at the same age. Second, we briefly investigate whether successful reproduction one year involves a fitness cost the following year. Third, we explore whether the sex ratio of kids produced by individual females suggests a bias in offspring sex ratio to maximize fitness returns. In later chapters we will argue that an understanding of female reproductive strategy is essential to understand mountain goat population dynamics, and that female reproductive strategy is also key to the conservation and management of this species.

Correlates and Consequences of Age of Primiparity

Many female mountain goats at Caw Ridge did not begin to reproduce until five years of age or later (fig. 7.1). The late age of primiparity limited population growth because females aged two to five years contributed much less to recruitment than in most other ungulate populations (Gaillard et al. 2000a). Because mountain goats do not appear to live longer than other ungulates (chapter 9), late primiparity shortened the average female's reproductive life span by two to three years compared to other species of comparable body mass. To examine what individual and environmental characteristics affected age of primiparity in

BOX 8.1

Measuring the Fitness Costs of Reproduction in Wild Ungulate Females

Although reproduction has both energy costs and fitness costs, only the latter affect the evolution of reproductive strategies. Energy costs are caused by the increased energy required to reproduce, including mating, gestation, lactation, and other forms of maternal care. Fitness costs, on the other hand, refer to a decrease in future reproductive potential caused by current reproduction. Gestation and lactation require substantial energy, but if forage resources are abundant, a female may simply obtain that energy through increased nutrition, without any effects on her fitness. Energy costs of reproduction that do not translate into fitness costs do not affect the selective pressure for restraint in reproductive effort. An iteroparous reproductive strategy, such as that of all female ungulates, implies that not all available resources are used for the current reproductive episode, so as not to compromise the ability to reproduce later. Therefore, it is fitness costs and not energy costs that shape the evolution of reproductive strategies.

A fundamental assumption of life-history theories is that resources are limited, therefore individuals cannot maximize all components of their lifetime reproduction. Instead, they should be forced into trade-offs in resource allocation, which should lead to fitness costs of reproduction. Current reproduction may compromise future reproduction, either by decreasing an individual's chance to reproduce again, by reducing its survival, or by leaving fewer resources available for future offspring (Stearns 1992). For female ungulates that reproduce many times over their life, natural selection should favor the best allocation of reproductive effort over the lifetime. In some cases the best reproductive strategy may be not to reproduce in one year, but rather accumulate resources, grow, and have a better chance at successful reproduction at some later time. The optimal lifetime allocation of reproductive effort will be affected by age-specific changes in mass, body condition, survival, and ability to cope with the energetic requirements of reproduction. The expected amount of variability in resource availability from year to year will also profoundly affect reproductive strategy (Stearns 1992).

To assess the fitness costs of reproduction in mountain goats, we used a correlational approach: we compared the reproductive success for females that had and had not reproduced the previous year. That is somewhat unsatisfactory because it cannot account for individual differences in ability to reproduce. Correlational approaches to fitness costs of reproduction have been criticized (Reznick 1985, 1992) because they assume that all individuals have equal reproductive potential. Typically, instead, individuals with greater reproductive success also have higher reproductive potential (Noordwijk and de Jong 1986). Individual differences can lead to positive correlations between

BOX 8.1
Continued

components of reproductive success, where life-history theory predicts negative correlations (Festa-Bianchet 1989a). When a measure of reproductive potential, such as individual mass, can be accounted for, individual differences in
reproductive potential can be partly accounted for. Festa-Bianchet et al.
(1998) measured the fitness costs of reproduction in bighorn ewes by taking
into account individual mass and population density. They found that weaning
a lamb one year decreased the chances of weaning a lamb the following year.
The fitness costs of reproduction increased with population density and decreased with ewe body mass: ewes with more resources (with large body mass,
or in years of low population density) suffered lower costs of reproduction
than ewes with fewer resources.

Because individuals vary in reproductive potential, the costs of reproduction cannot be accurately estimated by correlational studies (Reznick 1992).
One way to circumvent this problem is to artificially manipulate reproductive
effort, by randomly increasing or decreasing the number of juveniles reared by
different individuals. Numerous manipulations of reproductive effort have
been done with birds, where it is relatively easy to remove or add eggs to nests
(Moreno et al. 1997; Pettifor 1993; VanderWerf 1992). The reproductive effort of mammals, however, is more difficult to manipulate. Most experimental
studies of reproductive costs have been performed with rodents whose young
remain in a nest or burrow for several days or weeks (Hare and Murie 1992;
Humphries and Boutin 1996). Manipulation of litter size in wild ungulates is
almost impossible: females would not accept additional offspring. To manipulate reproductive expenditure one could prevent some females from reproducing with contraceptive implants (MacWhirter 1991) or remove neonates. Calf
harvest in the fall apparently increases subsequent reproductive success of female reindeer (Kojola and Helle 1993). The only published experimental manipulation of reproductive effort in free-ranging ungulate, in feral sheep, confirmed the presence of fitness costs of reproduction and suggested that costs
measured by phenotypic correlations are not a gross underestimate of those
obtained through experimental manipulation (Tavecchia et al. 2005).

Despite the likely influence of unmeasured individual differences in reproductive potential, our analyses of primiparity suggested a fitness cost because
kid production at five years of age was lowered by primiparity at four years.
Our approach is therefore very conservative and it is likely that the short-term
fitness costs of early primiparity would be even higher if differences in individual reproductive potential could be taken into account. Whether our inability
to detect fitness costs of reproduction in adults was due to individual differences in reproductive potential or to a strategy of low maternal investment remains to be seen.

females, we focused on population density and body mass during early development, two factors known to affect primiparity in other ungulates (Crête et al. 1993; Gaillard et al. 1992; Jorgenson et al. 1993a; Langvatn et al. 1996; Sæther and Heim 1993; Swihart et al. 1998). Studies of other ungulates suggested that primiparity occurs early for heavy females, and late at high population density, although in many cases most of the variation in age of primiparity was not explained by either or both of these variables.

Our analyses of primiparity had to consider carefully females caught and drugged at three or four years of age, because of the apparent negative effects of capture on reproduction the following year for these age classes (Côté et al. 1998a). We first compared body mass and horn length during early development for females that did and did not first reproduce at three or four years of age. This analysis excluded females drugged at age three, but included females drugged at four years because reproduction at age four was not affected by capture.

Our sample included three females that first reproduced at age three, twenty-six females that were primiparous at age four, and twenty-three that first reproduced at five years or later. Age-specific body mass and horn length data, however, were only available for subsets of this overall sample. Females that first reproduced as three or four year olds were on average 8% heavier as yearlings than those that postponed first reproduction until five years or later, but the differences in mass were barely significant (table 8.1). Surprisingly, for two year olds the trend was reversed: early reproducers were 3% lighter than late reproducers. There was no difference in horn growth during the first two years of life according to age of primiparity (table 8.1). Therefore there was no clear

TABLE 8.1

Mass as a Kid, Yearling, Two Year Old, and Adult, and Length of the First Horn Annulus for Female Mountain Goats That Did and Did Not First Reproduce by Four Years of Age

Variable	Early primiparae	Late primiparae	t	P
Mass as a kid	15.1 ± 1.73 (8)	14.3 ± 2.71 (7)	0.66	0.52
Mass as a yearling	33.3 ± 2.77 (14)	30.8 ± 2.69 (8)	2.48	0.055
Mass at 2 years	45.5 ± 2.88 (14)	46.7 ± 4.37 (12)	0.79	0.44
Adult mass (6 to 10 years)	70.9 ± 4.71 (15)	71.1 ± 6.21 (8)	0.07	0.94
First annulus	152.6 ± 10.6 (19)	151.8 ± 12.5 (13)	0.20	0.84

Mass (kg) on July 15 (mean ± SD) and length of the first horn annulus (mm) for females of Caw Ridge, 1989–2003. Sample sizes in parentheses.

evidence that early-reproducing females were heavier or in better condition than late-reproducing ones during early development.

For other ungulates of similar size as mountain goats, population density often has a strong effect on age of primiparity (Jorgenson et al. 1993a; Langvatn et al. 1996). Age of primiparity is one of the first vital rates to react to changes in population density: in some increasing populations, young females started delaying first reproduction before any other signs of density-dependence (such as lower juvenile survival) could be detected (Gaillard et al. 2000a). High sensitivity to population density (and, presumably, to resource availability) suggests that age of primiparity is an important and flexible component of a female's reproductive strategy (Festa-Bianchet et al. 1995). For female goats on Caw Ridge, however, the number of females in the year of birth had no effect on age of first reproduction,[4] possibly because population size only varied from thirty-five to fifty-four females over the period considered. Therefore, neither individual mass nor population density was strongly associated with variation in age of primiparity.

Because four-year-old females are lighter than older ones (chapter 6; see Côté and Festa-Bianchet 2001d), to wean a kid they may have to make a greater reproductive effort than what is required of older females, which could lead to a fitness cost of early reproduction. To examine that possibility, we compared the probability of producing a kid for five year olds that had and had not produced a kid as four year olds. If early primiparity led to fitness costs, females that were primiparous at four years should have been less likely to reproduce the following year compared to females that had not devoted any energy to reproduction at age four. We included in this comparison the three females primiparous at age three, and considered whether or not they produced a kid at age four. We excluded females drugged at age four because chemical immobilization may affect the probability of reproducing at age five (Côté et al. 1998a). This analysis revealed a substantial short-term fitness cost of early primiparity: five-year-old females were almost twice as likely to produce a kid if they had not reproduced the previous year (68%) than if they had reproduced (37%).[1] On the other hand, we found no evidence that early primiparity involved long-term fitness costs. For example, age of primiparity did not affect adult mass of females (table 8.1), suggesting that early breeders did not suffer a permanent growth cost.

The "decision" of when to first reproduce is a difficult one for female ungulates with a fixed and short breeding season. Most female goats that survive to three years of age will have only six to eight reproductive opportunities over their lifetime. One additional kid could be a major contribution to a female's fitness. A young female mountain goat, however, faces a number of risks and uncertainties over her first reproduction, which may make it difficult for natural selection to shape an optimal strategy. The average three-year-old female has accumulated only 80% of her asymptotic mass by the late November rut. Allocation of resources to gestation and lactation will likely reduce her body growth, at least over the short term. Because there is a single short rutting season each year, however, a female that does not conceive will not have another breeding opportunity until a year later. For a young female with enough body resources to reproduce successfully in an "average" year but not in a "difficult" year, the choice is complex. Whether or not her reproductive attempt will be successful and how it will affect her subsequent short-term reproductive potential depends upon several unpredictable variables. Weather over the next ten months and changes in population density or forage productivity may turn a successful reproductive event into an unsuccessful one that could also compromise her ability to reproduce the following year. On the other hand, unpredictable increases in predation risk (Festa-Bianchet et al. 2006) may select for higher propensity to breed. All these variables should affect the reproductive strategy of young females, which must make a greater reproductive effort than fully grown females (Clutton-Brock 1991). Consequently, young females should be more likely than mature ones to show evidence of reproductive costs. We found that many young females suffered a short-term cost of primiparity and had to forfeit reproduction the following year, presumably to recover body condition and avoid compromising their survival. Because ecological circumstances are very variable from year to year, females may adopt a reproductive strategy based on long-term "average" conditions, which may not be the best strategy in some or even in most years (Clutton-Brock et al. 1996). Although some young females appeared to pay short-term fitness costs of early primiparity, however, the risk of reproducing at four years of age was limited because early primiparity had no detectable effect on lifetime reproductive success. Clearly, a larger sample of individual females monitored over their entire life span, and a better measure of fitness than just reproductive success (Brodie and Janzen 1996; Coulson et al. 2003, 2006), are required to fully evaluate the costs and benefits of variability in age of first reproduction.

Fitness Costs of Reproduction and Adult Female Reproductive Strategy

Female ungulates in temperate environments generally sustain high energetic costs of reproduction, particularly for lactation. Lactating females tend to be lighter and have lower fat deposits at the onset of winter than nonreproducing females (Chan-McLeod et al. 1999; Festa-Bianchet et al. 1998; Green and Rothstein 1991).

Despite the methodological problems of using correlational evidence to measure reproductive costs (box 8.1), some studies have demonstrated the existence of fitness costs of reproduction in ungulates. Successful reproduction one year is often associated with a decrease in reproductive success the following year, or the cumulative costs of reproduction suggest a trade-off between different components of lifetime reproductive success (Bailey 1991; Bérubé et al. 1996; Birgersson 1998b; Clutton-Brock et al. 1983, 1996; Festa-Bianchet et al. 1998; Green 1990; Mysterud et al. 2002). In mountain goats, we have shown above that there is a short-term fitness cost of early primiparity. Primiparous female are a special case because their reproductive strategy must account for their incomplete somatic growth, in addition to balancing current and future reproduction.

For many ungulates, the fitness costs of reproduction become more evident when resources are scarce or when additional stressors, such as disease, affect a population. These patterns are to be expected if fitness costs arose through trade-offs in allocation of a limited resource, because when resources are abundant and animals are in good condition trade-offs may not be necessary, or may have limited importance. With few exceptions, when the fitness costs of reproduction are measurable, they are typically limited to a short-term reduction in reproductive potential. Negative effects of reproduction on adult female survival have only been shown in very high-density populations of ungulates on predator-free islands. In a food-limited population of red deer, lactation was associated with lower age-specific survival of hinds, particularly for very young and very old hinds (Clutton-Brock et al. 1983). In feral sheep, survival of lactating ewes was lower than that of nonlactating ewes in years of population crashes (Tavecchia et al. 2005). In bighorn sheep, survival costs of early primiparity were only evident during a pneumonia epizootic (Festa-Bianchet 1989a), while no survival costs of reproduction could be detected for adult or senescent ewes in two populations (Festa-Bianchet 1989a; Festa-Bianchet et al. 1998). In the Ram Mountain population of bighorn sheep, high population density increased both the fitness cost of weaning a lamb (Festa-Bianchet et al. 1998) and the additional cost of weaning a son over weaning a daughter (Bérubé et al. 1996). Other studies, however, did not consistently find the negative correlations between fitness components that are predicted by life-history theory, but rather reported that individual females that reproduced successfully in one year were able to reproduce again the following year. Such results were obtained, among others, in a long-term study of bighorn sheep at Sheep River (a population that appeared limited by disease and predation, not by food availability) and in an expanding Alpine ibex population reintroduced into vacant habitat (Festa-Bianchet 1989a; Toïgo et al. 2002). It is likely that stronger evidence of fitness costs of reproduction will be found in food-limited populations of ungulates, at densities where resource limitations are apparent.

The fitness cost of reproduction is difficult to measure reliably in wild ungulates. A large sample of marked individuals must be monitored over several years because fitness costs may not be evident when resources are abundant. In addition, individual differences in reproductive potential may hide fitness costs. Although we found a very weak association between female mass and reproductive success (chapter 7), it is reasonable to suspect that individual differences in reproductive performance are associated with differences in reproductive potential. Midsummer mass may not be an adequate estimator of reproductive potential in mountain goat females.

To examine the fitness costs of reproduction in adult females, we limited our analyses to those aged six to twelve years. We therefore avoided the complicating effects of trade-offs between growth and reproduction in young females and the effects of senescence in old ones. For individual females, kid production one year was independent of kid production the previous year. The probability of producing a kid was 76% following a year when the female had lactated (236 female-years) and 77% following a year when the female was not seen with a kid (83 female-years). Including in the analysis kid sex the previous year, number of adult females (a measure of population density), and female average adult mass did not produce any more insights into the costs of reproduction (logistic regressions, all $P > 0.17$). Weaning a kid had no effect on weaning success the following year (63% weaning success in 193 female-years following the weaning of a kid; 62% following 124 female-years when the female failed to wean a kid). Including kid sex, number of adult females and average female mass in statistical analyses did not provide any indication that weaning success lowered the probability of weaning a kid the following year (logistic regressions, all $P > 0.16$).

There was no measurable effect of reproduction on maternal survival. For 256 female-years of presenescent adult females (aged five to nine years), survival was identical at 95.3% following summers when the female did and did not wean a kid. For older females ($n = 79$ female-years), survival was 77.5% after weaning a kid and 76.7% following summers when they did not wean a kid. Therefore, we were unable to detect any short- or long-term fitness costs of reproduction for adult females.

In many sexually dimorphic ungulates, sons are costlier to wean than daughters. In bighorn sheep, weaning a son increases parasite counts and decreases weaning success the following year, compared to weaning a daughter, and as population density increases, the differential cost of sons also increases (Bérubé et al. 1996; Festa-Bianchet 1989a; Hogg et al. 1992). Greater fitness costs of sons than of daughters have also been reported in several species of deer (Birgersson 1998b; Clutton-Brock et al. 1983; Hewison and Gaillard 1999). In those species, however, sons are markedly heavier than daughters at weaning, while in mountain goats the effect of sex on kid mass is limited (chapters 6 and 7). Sex of the kid weaned one year had no effect on weaning success the following year. Weaning success was 60% (N = 96) after weaning a son, and 59% (N = 108) after weaning a daughter. In other ungulates, the greater fitness costs of sons compared to daughters are assumed to be related to greater fitness returns for mothers that make an additional investment in a son than in a daughter. Weaning a large son may increase the chances that it will become a dominant male and have high reproductive success by out-competing other males (Clutton-Brock et al. 1981). Although very little is known about the relationship between size at weaning and adult competitive ability in ungulates, it does appear that in some species juvenile development has a greater effect on adult size for males than for females (Festa-Bianchet et al. 2000; Hewison and Gaillard 1999). The lack of sex-differential reproductive costs and the very limited sexual dimorphism at weaning in mountain goats (chapter 6) suggest that differences in maternal care may have limited effects on the lifetime reproductive success of males in this species, or that the effects of maternal care on offspring fitness may not differ according to offspring sex.

The Conservative Reproductive Strategy of Mountain Goat Females

The only fitness cost of reproduction that we could detect for mountain goats was for primiparous females. We cannot predict what our analyses would have revealed had we been able to account for differences in individual reproductive potential. Recent theoretical and empirical studies suggest that reproductive trade-offs at the individual level may be less prevalent than expected, if genetic trade-offs occur at the level of resource acquisition rather than resource allocation (Dobson et al. 1999; Houle 1991; Kruuk et al. 2002; Reznick et al. 2000) or if individual differences are sufficiently strong to affect survival in addition to reproductive potential (Tavecchia et al. 2005). Regardless of how differences in individual reproductive potential or in ability to acquire resources may have affected the life history of individual females, however, successful reproduction one year did not compromise a female's ability to reproduce again the next year. We interpret that result as an indication that female mountain goats restrain their yearly reproductive effort.

Female mountain goats apparently adopt a very conservative reproductive strategy, similar to bighorn sheep (Festa-Bianchet and Jorgenson 1998). Female goats may have been selected to restrain their investment in any single reproductive event, possibly to avoid compromising their own survival. Based on the mass of kids at weaning relative to maternal mass, mountain goats appear to have a lower level of maternal effort compared to other ungulates (fig. 8.1). The offspring:mother mass ratio

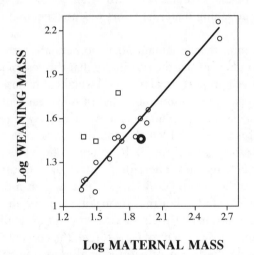

Log MATERNAL MASS

Figure 8.1. Maternal and offspring mass (kg, log-transformed) at weaning for temperate ungulates. The relationship for single births (circles) is Log (weaning mass) = 0.86 Log (maternal mass) − 0.039 and explains 96% of the variability in offspring mass. Squares refer to twin offspring, while data for mountain goats are indicated by a thicker symbol. When data were reported as carcass weight, an additional 30% was added to approximate live weight. We used our data for bighorn sheep, mountain goats, and white-tailed deer. Other sources were personal communications (J. Byers, pronghorn; T. Coulson, Soay sheep and red deer; S. Couturier, caribou; A. Loison, chamois; J.-M. Gaillard, roe deer) and the literature (European Mouflon, Garel et al. 2005; fallow deer, Birgersson 1998b; moose, Lynch et al. 1995; bison, Green and Rothstein 1991; elk, Boyce 1989).

at weaning (end of September) for mountain goats was about 0.36, compared to an average of 0.53 for other ungulates with maternal mass in the 45 to 100 kg range. Therefore, mountain goat mothers weaned offspring that were just over one-third their own mass, while in other ungulate species of similar size, offspring were weaned at just over half the mass of their mothers. The ratio was 0.42 for bighorn sheep, a species with a low level of maternal effort compared to other ungulates (Byers and Hogg 1995). If mountain goat mothers behaved like the average ungulate in terms of providing maternal care, the interspecific relationship in figure 8.1 suggests that at the end of September kids should have weighed 39 kg, about 10 kg (or 35%) more than their actual average mass (fig. 6.1). The interspecific comparison suggests that mountain goat mothers had a lower level of maternal effort than what may have been predicted based on their body mass. Because survival from birth to one year of age of mountain goat kids on Caw Ridge was higher than that reported in most other ungulate studies (Gaillard et al. 2000a), it appears that a high level of preweaning maternal effort was not required to ensure kid survival.

Before assuming that mountain goat mothers are extremely selfish, however, we should consider the possibility that the cost of producing an offspring of a given proportion of maternal weight may vary according to environmental conditions. Remember that primiparous four-year-old females suffered a cost of reproduction and were unlikely to raise a kid as a five year old. If four-year-old females and their kids were on figure 8.1, they would fall just slightly closer to the interspecific regression line than adult mountain goat females. Therefore, it may not be justified to assume that just because they wean small kids, mountain goat mothers make a minor reproductive effort: they may be making as large an effort as possible without compromising their survival and future ability to reproduce. In the harsh environment of Caw Ridge, the costs of reproduction per unit of maternal effort may be greater than in the habitats of the other ungulates in figure 8.1.

Based on the apparently low level of maternal effort, the fitness costs of reproduction in female mountain goats may also be lower than in most other ungulates. The low offspring:mother mass ratio at weaning is not surprising given the very long period of mass accumulation and body growth of mountain goats, especially males (chapter 6). A low level of maternal effort may also explain why there were no differences in the fitness costs of male and female kids. Male kids were slightly heavier than females at weaning (chapter 6), but still accounted for a lower proportion of maternal mass than what is common among most other ungulates.

The amount of maternal care received may play a limited role in the physical development of mountain goats, which are weaned when they have only reached about a third (for females) or a quarter (for males) of their asymptotic mass. As examined in later chapters, the low level of maternal effort and the long period of postweaning growth have important consequences for the population dynamics of mountain goats.

Kid Sex Ratio

Many studies of ungulates have examined the possibility that mothers vary offspring sex ratio according to their ability to provide maternal care (Sheldon and West 2004). Adaptive variation in offspring sex ratio is expected when the consequences of providing different levels of care vary according to offspring sex. Trivers and Willard (1973) suggested that in sexually dimorphic and polygynous mammals, females that can provide a large amount of maternal care should produce more sons than daughters. That is because the reproductive success of males is more variable than that of females and is more strongly affected by body or horn size. If maternal care affects adult size and therefore is correlated with offspring reproductive success, sons of high-quality mothers should have a much higher reproductive success than sons of low-quality mothers. Large males can monopolize many estrous females while small males should have very little or no success in competing for mates (Clutton-Brock et al. 1986; Coltman et al. 2002; Gomendio et al. 1990; McElligott et al. 2001). For daughters, the effects of body mass on reproductive success are likely smaller than for sons. Although no study has quantified how mass affects lifetime reproductive success for both sexes, differences in maternal care should have a smaller impact on the reproductive success of daughters than of sons (Clutton-Brock et al. 1984; Festa-Bianchet et al. 2000). In addition, low-quality mothers could benefit from producing daughters if sons had greater fitness costs, as reported in several species of ungulates (Bérubé et al. 1996; Clutton-Brock et al. 1981; Hewison and Gaillard 1999).

Evidence supporting the Trivers and Willard (1973) hypothesis has been provided by several studies of ungulates, but as reviewed by Hewison and Gaillard (1999) and Sheldon and West (2004), other studies found either no effect of maternal quality on offspring sex ratio or a trend opposite to that predicted by Trivers and Willard. For example, dominant red deer hinds at low population density produced more sons than daughters (Clutton-Brock et al. 1986) but dominant bighorn sheep ewes produced more daughters than sons (Festa-Bianchet 1991), and the relationship for

red deer disappeared at high density (Kruuk et al. 1999a). Different studies that compared offspring sex ratio to maternal condition or maternal
size reported trends in opposite directions for the same species: red deer
(Clutton-Brock et al. 1986; Post et al. 1999a) and roe deer (Hewison and
Gaillard 1996; Wauters et al. 1995). In bighorn sheep, a species that fits all
the Trivers-Willard assumptions (Hewison and Gaillard 1999) and is
therefore a suitable candidate to test their hypothesis, mothers do not bias
lamb sex ratio according to either their absolute body condition or their
body condition relative to the population mean (Blanchard et al. 2004).

Contradictory findings may be partly due to a publication bias: results
are easier to publish when they are significant and agree with currently
popular theories than when they are not significant or challenge accepted
wisdom (Festa-Bianchet 1996; Palmer 2000). A recent review of offspring
sex ratio bias in ungulates succeeded in justifying the exclusion of the
bighorn sheep data, demonstrating that even when results are published,
they can be excluded from comparisons if they challenge current dogma
(Sheldon and West 2004). There is substantial bias in studies of sex ratio
bias. The risk of publication bias is particularly strong for sex ratio data
because adaptive explanations exist for deviations of sex ratio from unity in
either direction. While the Trivers and Willard (1973) model predicts that
mothers in good condition should produce sons, the local resource competition hypothesis (Silk 1983) predicts the contrary: mothers in good
condition should produce daughters. Because males are the dispersing sex
in most mammals, mothers in poor condition should produce more sons
than daughters to reduce future competition within their home range.
Some studies of ungulates (Hewison and Gaillard 1996; Post et al. 1999a;
Verme 1985) have supported the local resource competition hypothesis,
but others have not (Clutton-Brock and Iason 1986).

Other factors can modify the adaptive value of producing offspring of
different sex. Life-history theory predicts that reproductive effort should
increase with age as reproductive value decreases (Stearns 1992). Therefore, offspring sex ratio may vary as females age, and one may expect an
increasing proportion of sons if the additional fitness cost of raising a
male was less important for older females. Rutberg (1986) reasoned that
females that had not reproduced the previous year should be in better
condition than those that had weaned an offspring and, according to the
Trivers and Willard hypothesis (1973), should produce more sons than
daughters. Finally, if sons are costlier than daughters, mothers that produced a male one year may be selected to avoid producing another male
the following year, as suggested for bighorn sheep and feral horses
(Bérubé et al. 1996; Monard et al. 1997).

In an earlier analysis of offspring sex ratio variation in the Caw Ridge mountain goats (Côté and Festa-Bianchet 2001c) we used generalized linear mixed models (GLMMs) to assess the effects of maternal age, social rank, and reproduction in the previous year on kid sex ratio. GLMMs allow random effects such as the identity of individual mothers that are repeatedly sampled to be examined within the framework of logistic regression (Steele and Hogg 2003). We found that the proportion of sons increased as mothers aged (fig. 8.2), while none of the other variables had significant effects. We now have six more years of data on kid sex ratio. Here, however, we used simpler comparisons examining one variable at a time, because our analyses confirmed that most variables had no effect on kid sex ratio.

For kids of marked mothers, overall sex ratio between 1989 and 2003 was even (167 males, 164 females). The proportion of daughters decreased with maternal age,[2] although the age effect was weaker than we previously reported based on data up to 1997 (Côté and Festa-Bianchet 2001c). The proportion of daughters decreased from 59% for mothers aged six years and younger to 38% for mothers ten years and older (fig. 8.3). These results could be due to an age-specific bias in offspring sex

Figure 8.2. Females produced more males as they aged. Here, #41 who produced six males in consecutive years from ten years old until fifteen years old. Photo by S. Côté.

ratio, but they could also arise if long-lived mothers had an age-independent tendency to produce more sons than daughters. Individual mothers monitored for a minimum of seven years ($n = 37$), however, produced on average 61% daughters when aged less than eight years but 46% daughters when eight years or older,[2] supporting the contention that mothers produced more sons as they aged.

Reproductive status did not affect kid sex ratio the following year. Females that had weaned a kid produced 51% sons ($n = 128$) while other females produced 52% sons ($n = 148$). Females that produced a son one year produced 59% sons the following year, whereas females that had weaned a daughter gave birth to 48% sons.[3] Therefore, unlike bighorn ewes, mountain goat mothers did not avoid producing sons in consecutive years. Limiting the analysis to females aged six to twelve years led to similar conclusions.

Reproductive effort and sex of the kid produced the previous year had no effect on kid sex ratio. The literature on ungulates provides examples of sex ratio bias and lack of bias for these two variables, but no clear pattern. Birgersson (1998b) found no effect of previous reproductive effort on offspring sex ratio in fallow deer. Rutberg (1986) claimed that bison cows that were barren the previous year produced more sons than daughters, but Shaw and Carter (1989) found that previous reproductive effort had no effect on offspring sex ratio in another bison population. Bérubé

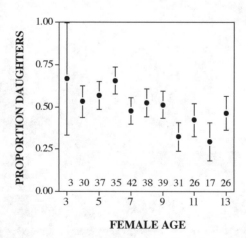

Figure 8.3. Proportion of female kids (± SE) produced according to maternal age of mountain goats on Caw Ridge, Alberta, 1989–2003. Sample sizes are shown for each age class at the bottom of the graph. Females thirteen years of age and older are pooled in a single age class.

BOX 8.2
Testing the Assumptions of the Trivers-Willard and Local Resource Competition Hypotheses

An unfortunate obstacle to the interpretation of sex ratio data is that competing theories provide adaptive explanations for results of opposite directions. The problem could be avoided if the assumptions of the competing theories were considered (Blanchard et al. 2004; Hewison and Gaillard 1999). The majority of studies that have addressed variation in offspring sex ratio in mammals, however, have not tested the assumptions of either model. A species that does not meet a model's assumptions is clearly not a good candidate to test that model.

Mountain goats do not meet the main assumption of the local resource competition hypothesis, that most young males (and no young females) should emigrate from the maternal home range. Fewer than a quarter of young mountain goats disperse, and both sexes may disperse. In addition, because all "nursery" goats share a common home range, an emigrant would not reduce forage competition for its mother more than for any other goat in the population. Any costs of producing an offspring of the dispersing sex (or of encouraging an offspring to emigrate) would be borne only by the mother, but the benefit of reduced competition would be shared with all other members of the population. Under those circumstances, there would not be any fitness benefit for a female that produced a son in order to decrease competition for forage in future years. Therefore, there is no justification to test the local resource competition hypothesis in mountain goats. The hypothesis may be relevant only for territorial species, and it would be interesting to test it in Japanese serow, where both sexes defend individual territories against conspecifics of the same sex (Ochiai and Susaki 2002).

The Trivers and Willard model makes three assumptions. First, it assumes that high-quality mothers will wean high-quality offspring. In mountain goats, maternal mass explained 17% of variability in kid mass (Côté and Festa-Bianchet 2001d). Therefore, mountain goats satisfy, barely, the first assumption of the Trivers and Willard hypothesis. A second assumption is that offspring that receive more maternal care will be of higher quality as adults than offspring that receive less care. We cannot currently test this assumption for mountain goats because few were weighed both as kids and as adults. Correlations between adult quality and juvenile quality, however, have been documented in other polygynous and sexually dimorphic ungulates (Birgersson and Ekvall 1997; Festa-Bianchet et al. 2000; Green and Rothstein 1991), and although many of these relationships are weak, they are generally positive and significant. Therefore, it seems reasonable to assume that preweaning maternal care in mountain goats has a positive effect on adult size and reproductive success of offspring.

BOX 8.2
Continued

The third assumption of the Trivers-Willard model is that adult pheno-typic quality has a greater influence on reproductive success of males than of females. This is a key assumption because it justifies the expectation of greater maternal care to sons than daughters. We cannot test this assumption for mountain goats. Among ungulates, this assumption has been adequately tested (and validated) only in red deer (Clutton-Brock et al. 1986). Hewison and Gaillard (1999) suggested that this assumption is probably valid in most polygynous ungulates where male mating success increases with body size and rank. Mountain goats may satisfy this assumption because male reproductive success is probably related to body size and social rank. Of the three key as-sumptions of the Trivers-Willard model, we can only directly test one and we must rely on evidence from other species to support the other two. We have no evidence, however, that any of these assumptions is not valid.

et al. (1996) found that bighorn ewes in two populations avoided produc-ing sons in consecutive years, but studies of other sexually dimorphic un-gulates did not reveal a similar pattern (Clutton-Brock and Iason 1986). Therefore, either the published studies happened to find sex ratio effects by chance, or sex ratio variation depends on several different variables, many of which remain to be identified, accounting for the contrasting re-sults obtained by different studies (Blanchard 2002).

As mountain goat females aged, they produced more sons (fig. 8.3). Similar results have been reported for other ungulates but, again, the pat-tern is not universal and its adaptive significance is the subject of debate (Hewison et al. 2002; Saltz and Kotler 2003). In one population of roe deer (Wauters et al. 1995) and one captive population of Cuvier's gazelle (Alados and Escos 1994), the proportion of female embryos decreased with maternal age. In barren-ground caribou, mothers aged up to four years produced 38% males and those older than ten years produced 67% males (Thomas et al. 1989). Kojola and Eloranta (1989), however, found no relationship between maternal age and birth sex ratio in reindeer. In very old bighorn ewes, the proportion of male lambs appears to decrease, but that may be because senescent ewes are in such poor condition that they are unable to care for sons (Bérubé 1997).

Why should mountain goat mothers produce more sons as they age? In sexually dimorphic ungulates, males are generally costlier to wean than females (Birgersson 1998a; Clutton-Brock et al. 1981; Gomendio et al. 1990; Hogg et al. 1992; Monard et al. 1997). For mountain goats, however, we found no evidence that sons were costlier than daughters and there was little sexual dimorphism in kid mass. Mountain goat females increase in body mass until about six years of age and presumably gain maternal experience with each reproduction; therefore, their ability to provide maternal care may increase with age. Older females are socially dominant (Côté 2000), but dominance rank has little effect on the reproductive success of mature females (Côté and Festa-Bianchet 2001d). Compared to young mothers, old mothers may either increase reproductive investment or provide greater maternal care without necessarily investing more (Cameron et al. 2000). If older mothers are better mothers, then our results support the Trivers and Willard model. Kid survival, however, did not vary with maternal age, and we cannot determine whether older females derive a fitness benefit by producing more sons. Development of sexual dimorphism in mountain goats is slower than in many other ungulates and occurs almost entirely postweaning (chapter 6). Our results, therefore, do not firmly establish an adaptive link between maternal age and offspring sex ratio. While we caution against assuming that a sex ratio bias is evidence of differential maternal investment in the sexes, our results suggest that female age structure in mountain goat populations may affect the sex ratio of the kid crop, with obvious implications for management strategies aimed at harvesting males (chapter 12).

Summary

- Age of primiparity in mountain goats appeared largely independent of individual body mass during early ontogeny and of changes in population density.
- Early primiparity carried a fitness cost, since five-year-old females were less likely to reproduce if they had produced a kid as four year olds. There was no evidence of long-term fitness costs of early primiparity.
- We found no measurable fitness costs of reproduction for adult females. The level of reproductive effort of adult mountain goat females appeared lower than for most other ungulates, but a lower level of maternal effort may represent a greater maternal investment in the harsh environment of Caw Ridge than in more productive habitats.
- Kid sex ratio was independent of maternal mass, previous reproductive success or sex of kid produced the previous year. As

females aged, they produced more sons, possibly supporting the Trivers and Willard model. The adaptive significance of changes in offspring sex ratio with maternal age, however, remains unclear.

Statistical Notes

1. G-test comparing the probability of producing a kid for five year olds that had and had not produced a kid the previous year, including kid production at four years for three females primiparous at age three. Females drugged at four years of age were excluded. $G = 4.41$, $n = 35$, $P = 0.036$.

2. Logistic regression of the proportion of sons with maternal age, $b = 0.095$, $\chi^2 = 5.836$, $P = 0.016$; $n = 324$). Wilcoxon signed-ranks test of the proportion of daughters by thirty-seven females (monitored for a minimum of seven years) when aged less than eight years (61%) and when eight years or older (46%), $z = 2.03$, $P = 0.042$.

3. Females that produced a son one year produced 59% sons ($n = 70$) the following year, females that had weaned a daughter produced 48% sons ($n = 84$); $G = 1.842$, $P = 0.20$.

4. For forty-one females born at Caw Ridge from 1989 to 1999 and not drugged at three years of age, logistic regression comparing primiparity at three to four years or at five years or older with the number of females in the year of birth: $P = 0.68$.

CHAPTER 9

Survival and Dispersal

Survival of mountain goats varies substantially according to age, and adult male mortality is much greater than the mortality of adult females. Here we will describe survival senescence and report on the maximum lifespan of both sexes. We conclude by examining the impacts of emigration and immigration on the size and sex–age structure of the population.

Recent studies have underlined the importance of temporal changes in sex and age structure in the population dynamics of ungulates, including mountain goats (Coulson et al. 2001; Festa-Bianchet et al. 2003; Gaillard et al. 2001). The sex–age structure of our study population changed over time, as is typical of many unmanaged ungulate populations. Because mortality and reproductive success vary substantially among sex–age classes, changes in sex–age structure can lead to large differences in population growth rate, independently of resource availability or of other extrinsic factors. The effects of sex and age on survival and reproduction may be stronger for mountain goats than for other ungulates, because local survival (i.e., persistence in the population) was very different according to sex, while age had profound effects on both survival and reproduction. As we saw in chapter 7, females aged two to four years contribute little to recruitment, therefore differences in female age distribution among years will affect the productivity of the population. In addition, juvenile ungulates are much more susceptible than adults to the effects of both density and weather (Gaillard et al. 1998), so that the growth rate of a population with a large proportion of juveniles should be affected by changes in density or in weather much more than that of a population with a small proportion of juveniles (Coulson et al. 2001). In particular, the survival of prime-aged adult females appears mostly unaffected by changes in density or in environmental conditions in most ungulate populations (Gaillard et al. 1998).

Cause of Death

We know little about the causes of mortality. In the first four years of the study, we sought to identify causes of kid mortality by attempting to locate radio-collared kids each day. At other times during the study and for other sex–age classes, however, we seldom knew the cause of death of marked goats, or even if the goat had died or emigrated. For over 80% of marked goats, our monitoring ended with a final sighting (often at the end of a field season), then the animal simply disappeared. We know that a few emigrated, because they were seen or radio-tracked elsewhere. Therefore, our data on survival reflect "apparent survival" (Lebreton et al. 1992), or "local survival," because we could rarely distinguish mortality from emigration. Because emigration appeared mostly restricted to goats aged two to four years, however, disappearances of other age classes can be assumed to be mainly due to mortality. Although radio collars allowed us to find carcasses, many goats died during winter. Typically, by the following spring all we found was the radio and bones, so it was difficult to establish cause of death. The sparse results we obtained suggest that predation was the main cause of mortality (table 9.1). Goat 105 (table 9.1) illustrates the problems we faced in determining cause of death. She was alive and apparently healthy one evening, and was dead the following morning. She had no external injuries and we do not know why she died. Had we not recovered her carcass for another day or two, she would have been scavenged and we may have attributed her death to predation. Goat 132 was alive and healthy one evening, and fed upon by a grizzly bear by the following morning. We assume the bear killed it, but cannot exclude that it died and the bear scavenged it. In both cases, less than fifteen hours elapsed between our last sighting of the goat alive and our first sighting after its death, yet we could not be certain of why each goat had died.

The apparently large proportion of deaths due to research activities (table 9.1) overestimates the impact of our research program on mortality, because cause of death was unknown in an overwhelming majority of cases. We always knew when a goat died during trapping, but we rarely knew what caused natural deaths. Trapping accidents accounted for about 3% of deaths and were excluded from calculations of sex- and age-specific mortality.

TABLE 9.1

**Known and Suspected Cause of Death of Marked Mountain Goats,
1989–2004**

ID	Sex	Age at death	Year	Cause of death or disappearance
19	M	Kid	1989	Wolf predation
33	M	Kid	1989	Wolf predation
42	F	9+	1992	Wolf predation (?)
58	M	Kid	1990	Wolf predation
79	F	Kid	1992	Wolf predation (?)
94	F	12	1999	Wolf predation
150	F	1	1995	Wolf predation
172	M	Kid	1996	Wolf predation
43	M	Kid	1990	Unknown predator (?)
44	F	Kid	1989	Grizzly bear predation
61	M	1	1991	Grizzly bear predation
74	M	Kid	1990	Grizzly bear predation
76	F	1	1991	Grizzly bear predation
69	F	Kid	1990	Grizzly bear predation
91	F	Kid	1992	Grizzly bear predation
132	M	3	1996	Grizzly bear predation
52	M	5	1992	Cougar predation (?)
66	F	Kid	1991	Cougar predation
70	F	1	1991	Cougar predation
87	F	Kid	1991	Cougar predation
291	F	3	2004	Cougar predation
72	F	Kid	1990	Fall from cliff
45	F	4	1993	Capture-related
81	F	Kid	1991	Capture-related
95	F	3	1995	Capture-related
101	M	1	1993	Capture-related
237	M	2	2000	Capture-related
245	M	Kid	2000	Killed in trap by another goat
186	F	2 days	1997	Neonatal mortality
247	M	<1 day	2000	Neonatal mortality: born alive but never suckled
105	F	11	2003	Digestive problem (?)

From Caw Ridge, Alberta. Grouped by categories of causes of death or disappearance. Question marks refer to cases where we could not be certain of the cause of death.

Kid Survival

Mammals do not disperse before weaning, therefore we can assume that all kid disappearances were due to mortality. We assessed survival of kids in two ways: by comparing the total number born each year to the total surviving to weaning and to one year, and by monitoring the survival of known individuals. Some kids were known because they were marked, others were recognizable through their association with marked mothers. We knew the sex of all marked kids and of most of those recognized because they associated with marked mothers. Therefore many unmarked kids were included in the sample we used to measure sex-specific survival.

Studies of ungulates that rely on observations of mother–offspring pairs to assess juvenile production and survival typically cannot distinguish neonatal mortality from cases where individual females do not reproduce. For example, in the Ram Mountain bighorn sheep population, as many as half of the lambs born in a given year may die at birth (Portier et al. 1998) and are detected only because females are regularly recaptured to check their lactation status. For mountain goats at Caw Ridge, however, we suspect that undetected neonatal mortality was less than 1%. Over the entire study, we only saw or captured two adult females that showed signs of lactation but were never seen with a kid. In addition, females normally isolate themselves a few days before giving birth, and it can be assumed that females that remain in nursery groups during the parturition season do not give birth. During late-May censuses between 1994–1999 and 2002–2003, only one adult female was observed alone but never seen with a kid. All other females either remained within nursery groups (suggesting that they did not give birth) or were seen with a kid (including two kids that were dead by the time we saw them).

Survival from birth to one year during our study was higher than reported for most other ungulates. Of 355 kids born from 1989 to 2002, 87% survived to weaning, and 64% survived to one year of age. Gaillard et al. (2000a) reviewed studies of forty-four populations of twenty-three species of ungulates and reported an average survival to one year of 50% (SD = 20.2, range 14–97%). In 70% of those studies, survival to one year was less than at Caw Ridge. There are few estimates of kid survival for other populations of mountain goats, and most rely on changes in kid:female ratios rather than on monitoring marked kids or on complete total counts. Age ratios are unreliable to assess juvenile survival in mountain goats because of the substantial but variable proportion of young females that do not reproduce, and because often males and females cannot be distinguished, forcing researchers to compare kid:adult ratios rather than kid:female ratios. The very limited data available on the reproductive success of marked individuals suggest that high kid survival to one year may be typical of mountain goats: it was estimated at 56% over four years in an introduced population in Colorado (range 46–78%) (Adams and Bailey 1982) and at 69% in a native population in Montana (Smith 1976). Similarly to juvenile survival in other ungulates (Gaillard et al. 2000a), however, the most striking feature of kid survival in mountain goats was not its average but its variability from year to year (fig. 9.1): it ranged from 38 to 92% over fourteen years and its coefficient of variation was 24.4%. Other studies of ungulates report very strong cohort effects: juveniles in some cohorts perform consistently better than those born in other years, with important consequences for population dynamics (Beckerman et al. 2002; Gaillard et al. 1997; Rose et al. 1998). If there were strong cohort effects in mountain goats, we should have recorded a correlation between preweaning and postweaning survival of juveniles born in the same year, as reported for the introduced population in Olympic National Park (Stevens 1983). Survival from birth to weaning and from weaning to one year, however, was not correlated within 14 cohorts,[1] suggesting that there were no strong cohort effects.

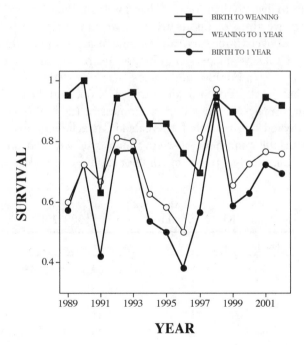

Figure 9.1. Survival of mountain goat kids born from 1989 to 2002 at Caw Ridge, Alberta, calculated from total population counts. Survival to weaning is from birth to August for 1989 to 1992 and from birth to September thereafter.

There were no differences in kid survival according to sex (Côté and Festa-Bianchet 2001a). Although survival to weaning between 1989 and 2003 was greater for males (91.6%, $N = 166$) than for females (84.6%, $N = 162$),[2] survival from birth to one year was almost identical: 67.1% for 146 males and 66.2% for 142 females. These figures are an overestimate, because only kids that survived to a few weeks or months of age were of known sex. This overestimate, however, is very minor because early mortality was rare and should not be affected by sex. Comparing these estimates of survival with those obtained by comparing total counts at birth, weaning, and one year of age suggests an overestimation of 1% for survival to weaning and 3% for survival to one year. Although we cannot exclude a possible sex bias in neonatal mortality, our data strongly suggest that kid survival was independent of sex. That is not unexpected given the very small difference in size between male and female kids (chapter 6).

Many studies of ungulates report a positive effect of neonatal mass on juvenile survival. Newborn ungulates, once located, are generally easy to capture by hand, but few studies have obtained data on mass of older juveniles. We did not attempt to catch newborn kids in the field because we did not want to disturb them or risk abandonment, therefore we have no data on birth weight. For bighorn lambs, weaning mass was positively correlated with overwinter survival, but only at high population density, and lamb survival was independent of mass at three weeks of age (Festa-Bianchet et al. 1997). For mountain goats, mass adjusted to mid-July had a positive but weak effect on kid survival (Côté and Festa-Bianchet 2001a). Our reanalysis with a larger sample size (table 9.2) confirmed that mass affected survival only for female kids. Excluding kids that were abandoned by their mothers, surviving male kids weighed on average 0.9 kg (or 7%) more than those that did not survive to one year, while fe-

TABLE 9.2
Mass and Kid Survival to One Year, 1989–2003

Sex	Survival	N	Mass	t	P
Male	Yes	26	14.0 ± 2.64	0.934	0.178
	No	13	13.1 ± 3.10		
Female	Yes	33	13.9 ± 2.56	1.894	0.033
	No	13	12.4 ± 1.91		
Both	Yes	59	13.9 ± 2.58	1.949	0.027
	No	26	12.7 ± 2.56		

Mass (kg ± SD) adjusted to July 15 for mountain goat kids of each sex at Caw Ridge, Alberta. One-tailed P-values are reported.

male kids that survived to 1 year weighed about 1.5 kg more than those that died, a 12% difference.

Yearling Survival

Sex-specific survival of yearlings was calculated for marked individuals only, because of the potential immigration of unmarked two year olds. Although one female that was last seen as a yearling left the study area (table 9.3), all other goats known to emigrate from Caw Ridge did so when aged two years or older (see below). Therefore we assume that almost all yearlings that disappeared had died. From 1989 to 2003, survival to two years was 73.5% for sixty-eight yearling males and 84.7% for seventy-two yearling females. The 11% sex difference is nearly significant.[3] Our results are similar to the 71% survival for seven radio-collared yearlings of both sexes reported by Smith (1986) in Alaska. No other estimates of yearling survival exist for mountain goats, and few studies have measured yearling survival of other ungulates. For twelve studies of nine species of ungulates, Gaillard et al. (2000a) reported an average yearling survival of 87% (SD = 8%, range 70–98%). Therefore, yearling survival in mountain goats appears lower than in most other wild ungulates. There are no strong sexual differences in yearling survival in other ungulates, although in some species yearling females appear to have greater survival than yearling males. In red deer and feral sheep, yearling males have lower survival than yearling females, especially at high density (Clutton-Brock et al. 1997a). Yearling males have higher mortality than yearling females also in Pyrenean chamois, a rupicaprid with little sexual dimorphism (Loison et al. 1999a). Yearling survival is independent of sex in the highly dimorphic bighorn sheep and in the weakly dimorphic roe deer (Loison et al. 1999a).

We did not mark enough yearlings to reliably compare variability in survival between years, especially given that the sexes had to be considered separately because of the higher survival of females. Between 1990 and 2002, we had fewer than ten marked yearling females in all but one year, and never more than eight tagged yearling males. Therefore, despite the possible immigration of unmarked two year olds, to assess yearly variability in survival from one to two years we relied on comparing the sex-specific total count of yearlings in June to the total count of two year olds the following year. Those calculations suggested an average yearling female survival of 84.8%, almost identical to that estimated from marked individuals. Over thirteen years, the coefficient of variation (CV) in yearling female survival was 20.6%, slightly less than for kid survival.

TABLE 9.3
Known and Suspected Emigration from Caw Ridge by Marked
Mountain Goats, 1989–2003

ID	Sex	Age	Month	Evidence
67	F	2	August	Seen elsewhere, left soon after being captured
92	M	2	winter	Radio signal 26 km SW of Caw Ridge 2.5 years later
98	M	1	winter	Remains found 110 km ESE of Caw Ridge
100	F	2	October	Seen about 8 km NE of Caw Ridge, with goat 133
133	F	1	October	Seen about 8 km NE of Caw Ridge, with goat 100
166	M	2	August	Radio signal on "mortality" mode 2 years after disappearance, 12 km from Caw Ridge
184	M	3	July	Radio signal on "mortality" mode 1 year after disappearance, 35 km from Caw Ridge
190	M	2	August	Radio signal on "live" mode 2 years after disappearance, 27 km from Caw Ridge
191	M	2	winter	Radio signal 3 km from Caw Ridge, goat returned the following July 24 with an unmarked male aged 2 years
208	M	2	August	Disappeared while wearing a functioning radio collar
224	M	2	September	Radio signal NE of Caw Ridge in early September 2000, returned by the following June, left again in June 2002, back on Caw Ridge in June 2003
249	M	4	June	Disappeared while wearing a functioning radio collar
280	M	2	August	Radio signal on "live" mode, 20 km from Caw Ridge

"Age" refers to the age of the goat when it emigrated, and "month" is the suspected month of emigration. For goats that were last seen on Caw Ridge at the end of a field season (September), we indicated "winter" as the month of emigration. Most radio collars switched to mortality mode (a higher rate of signal transmission) after being immobile for several hours.

For males, there was one obvious immigrant one year (there were two yearlings in 1992, and three two year olds in 1993). We assigned a survival rate of 100% to yearling males in 1992. Total counts suggested an average yearling male survival of 75.5% with a CV of 26.7%. Therefore, survival of yearling males appeared to be both lower and more variable than that of yearling females.

BOX 9.1
Estimating Survival of Adult Mountain Goats

The survival of goats older than one year was analyzed with a capture–mark–recapture technique using the program SURGE (Lebreton et al. 1992). This technique estimates the probability that an individual will be sighted if it is alive, based on how often individuals not seen one year are sighted in subsequent years. Because the probability of sighting an animal may vary according to its age and sex, and because in most studies not all surviving marked animals are seen every year, accounting for differences in sightability is extremely important (Boulinier et al. 1997). If differences in sighting probabilities are ignored, they may lead to spurious estimates of survival probabilities. In our case, however, sighting probability was not an important variable, as virtually all marked goats were seen every year. Sighting probabilities were 99.6% for females and 99.0% for males, almost entirely because of a few emigrants seen elsewhere after they had left. Three marked young males were not seen one year and were later sighted in the study area, but we suspect that they had left the ridge for one year and then returned. Most marked females were seen at least fifteen times and most males at least five times each summer.

Adult Survival

Local survival of adult mountain goats varied substantially according to sex (fig. 9.2). Males showed a very high rate of disappearance as two and three year olds, so that only 39% of yearling males were still on Caw Ridge as four year olds (fig. 9.3). Some young males emigrated (table 9.3), therefore we measured "disappearance" rather than "death." The yearly survival of males aged four to seven years was also lower than that of females of the same age. The cumulative effect of yearly differences in sex-specific mortality over these four years was substantial: of goats aged four years, our data suggest that 83% of females but only 63% of males would have survived to eight years. After eight years of age, males experienced much higher mortality than females. Less than 10% of yearling males but over 50% of yearling females survived to 10 years. Mortality of males aged eight to ten years was higher than that reported for males of the same age in seven of eight other ungulate populations with detailed data on sex- and age-specific survival, the exception being one population of roe deer (Catchpole et al. 2004; Loison et al. 1999a; McElligott et al. 2002; Toïgo et al. 1997). We know very little about the causes of death of males aged four years and older (table 9.1), partly because fourteen of

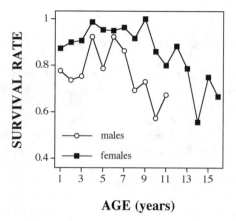

Figure 9.2. Age-specific survival of mountain goats on Caw Ridge, 1989 to 2003. One male that survived past eleven years of age died as a fifteen year old, and one female that survived past age sixteen died as an eighteen year old.

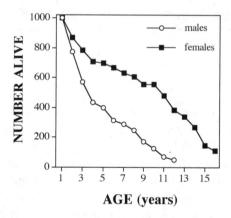

Figure 9.3. Persistence in the Caw Ridge population of an imaginary cohort of one thousand yearling mountain goats of each sex, calculated from the age-specific survival from 1989 to 2003.

twenty-three (61%) that disappeared did so between late September and early May, while we were not in the study area. Very few male mountain goats survive to old age: we marked 145 males, but only monitored 10 that survived to ten years of age.

The age-specific survival of mountain goat females was broadly similar to that of other ungulates (Benton et al. 1995; Loison et al. 1999a; Toïgo et al. 1997), except for the low survival of two year olds. Females of most other species show an increase in survival from yearlings to two year olds, then survival remains very high and stable from two to about seven or eight years of age (Gaillard et al. 2000a). The yearly survival of female mountain goats, however, continued to increase gradually from one to three years of age.

The survival of two-year-old mountain goat females appeared somewhat lower than for other ungulates: at 90%, it was similar to the survival of two-year-old female roe deer at Trois-Fontaines, France, but about 3 to 10% lower than the survival of two-year-old female roe deer, bighorn sheep, and red deer in five other study areas, which averaged 96% (Benton et al. 1995; Loison et al. 1999a). At least two two-year-old females emigrated (table 9.1), therefore the apparently low survival of this age class may partly be due to dispersal. Smith (1976) reported 100% survival for seventeen radio-collared two-year-old mountain goats of both sexes, possibly because by using aerial telemetry he was able to follow dispersing individuals. It is possible, however, that two-year-old females on Caw Ridge have lower survival than other female ungulates of the same age because they have not reached the same proportion of asymptotic body mass (chapter 6). In bighorn sheep, survival increased with body mass for lambs and yearlings but not for adults (Festa-Bianchet et al. 1997). Unfortunately, we did not have sufficient data on body mass for yearling and two-year-old goats to compare mass of those that survived and those that died. Because two-year-old females need to gain a substantial amount of mass, they may suffer a slightly higher mortality than females aged three to nine years.

Because of the differences in age-specific body growth, two-year-old mountain goats cannot be compared directly to two year olds of most other ungulate species. In roe deer, bighorn sheep, red deer, and chamois, two-year-old females are much closer to their asymptotic body size than in mountain goats. In several of those other species, two-year-old female may be parous, while at Caw Ridge it was exceptional even for three year olds to reproduce. Therefore, at Caw Ridge two-year-old females should be considered juveniles: they only weigh about 60% as much as fully grown adults, some are still associated with their mothers (chapter 5), and the vast majority are still two or more years ahead of their first reproduction. In red deer on Rum, the survival of two-year-old females was also about 90%, higher than for yearling females but lower than for females aged three to nine years (Catchpole et al. 2004). Red deer females on Rum are typically primiparous at three or four years of age and it is likely that two year olds are still growing in both body mass and skeletal size. We suggest that both the late age of primiparity and the lower survival of two-year-old female goats are consequences of their protracted period of body growth. Growing animals appear susceptible to various sources of mortality because they face a trade-off in allocating resources to body growth and to maintenance (Clutton-Brock et al. 1985).

Recent reviews of long-term studies of marked ungulates (Gaillard et al. 1998; Gaillard et al. 2000a) suggest that the average yearly survival of adult females, before the onset of senescence, is about 92%, ranging from 81% in one population of mule deer in Colorado to 98% in pronghorn antelope fenced within the National Bison Range in Montana. In that review, adult females aged two to seven years were considered presenescent (or prime-age), based on results from bighorn sheep, roe deer, and chamois that suggested that the onset of survival senescence was at about eight years of age (Loison et al. 1999a). On Caw Ridge, the survival of mountain goat females aged two to seven years was 94% (fig. 9.2), very similar to that reported for other ungulates. The prime-age period in mountain goat females, however, may extend to nine years (fig. 9.2).

Survival of females aged ten to sixteen years was clearly affected by senescence and averaged only 76%. Comparisons of survival of old females in different species or populations are difficult because of two reasons. First, sample sizes are small, even for long-term studies. Second, for animals older than about twelve years survival rates decline rapidly with age (fig. 9.2), therefore the exact age structure of a sample of "old" females has a strong influence on its survival. In one population of bighorn sheep and one of roe deer, survival of females aged ten to sixteen years was comparable to or lower than that of mountain goats on Caw Ridge. On the other hand, in another population of bighorn sheep, one of roe deer, one of red deer, and one of isard, survival of females aged ten to sixteen was higher than at Caw Ridge (Benton et al. 1995; Loison et al. 1999a). Overall, therefore, the survival of both prime-aged and senescent adult females on Caw Ridge appears similar to that reported for females of other species of ungulates.

Data on age-specific survival of mountain goats are available for only one other population, studied by Smith (1986) in coastal Alaska. He monitored twenty-four males and thirty-eight females, but did not present sex-specific survival. Excluding hunting mortality, survival of goats aged two to eight years averaged about 99%, based on 152 goat-years, but decreased to 68% for goats nine years of age and older, based on 25 goat-years. Therefore, the mountain goats studied by Smith (1986) had similar senescent survival to that of the Caw Ridge goats but enjoyed higher prime-aged survival. Even though Smith did not separate male and female survival, the 1% mortality rate that he recorded for prime-aged adults leaves little opportunity for sexual differences. The fact that the population was hunted, however, leaves open the possibility that the sample of males mostly included very young goats, if hunters selectively harvested males rather than females. Because he used radio telemetry, Smith (1986) was also able to monitor the survival of dispersing individuals.

Survival of females aged two to nine years did not vary much from one year to another, ranging from 86 to 100% with a CV of only 4.7% (fig. 9.4). The survival of males aged two to seven years, on the other hand, ranged from 60 to 94% with a CV of 13% (fig. 9.4).

Because the proportion of prime-aged females declined from almost 90% to about 60% as the population increased (fig. 9.5), adult female survival also decreased, simply because of changes in age structure. As pointed out by Festa-Bianchet et al. (2003), the average age of females typically increases with population density in ungulates, leading to a spurious correlation of population density and adult female survival. High-density unhunted populations often include many senescent females that have higher mortality than prime-aged females. If one ignores the effects of female age on survival and the positive correlation between population

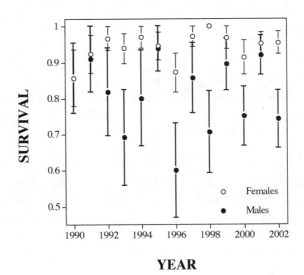

YEAR

Figure 9.4. Yearly variation in survival (mean ± SE) of mountain goat females aged two to nine years and males aged from two to seven years at Caw Ridge, 1990 to 2003. Yearly sample sizes averaged thirty-two for females and fifteen for males.

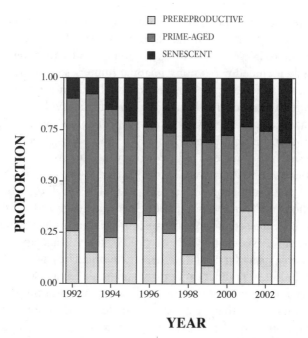

Figure 9.5. Age structure of adult female mountain goats at Caw Ridge from 1992 to 2003. Prereproductive females were aged two or three years, prime-aged females were between four and nine years, and senescent females were aged ten years or older.

density and average female age, the higher female mortality seen at high density could be attributed to density-dependence in adult female survival, while in reality it is due to changes in age structure.

Longevity

The oldest goats we monitored on Caw Ridge were a fifteen-year-old male and an eighteen-year-old female. Cowan and McCrory (1970) also reported an eighteen-year-old female. Many studies of ungulates report maximum longevities of fifteen to sixteen years for males and eighteen to twenty years for females (Benton et al. 1995; Hoefs 1991; Loison et al. 1999a).

The age of the "oldest known" individual, however, is an anecdote with little biological value. It is misleading to use the oldest known individuals to compare longevity among species or populations, because those data reveal little about the distribution of life span among individuals (Gaillard et al. 1994). It is more informative to compare the "rate of senescence," the decline in survival experienced by old individuals. Research on rupicaprins has an advantage in this respect, because horn annuli provide a much more reliable estimate of age for these species than for other ungulates, particularly for females (Loison et al. 1994; Miura and Yasui 1985; Stevens and Houston 1989). Age determination by counting horn annuli of female wild sheep, for example, is imprecise after about four years of age (Geist 1966; Hoefs and König 1984). For cervids, it is impossible to estimate precisely the age of either sex beyond about three to four years, after all milk teeth have been replaced by permanent teeth. We assumed that our age assignments were accurate for goats first caught when seven years of age or less, because some of our known-age animals did not form a clear annulus every year past that age (Côté 1999). Of the marked goats in our "known-age" sample, 90% were first caught when aged less than three years. Because we could reliably age individuals up to seven years of age, we could start monitoring the survival of "old" goats from within a few years of working at Caw Ridge, while most other studies of ungulates cannot begin to gather information on old individuals until after ten to twelve years of monitoring, when animals first caught as juveniles or yearlings (and therefore of known age) reach senescence. Because few wild ungulates survive to old age, and because the age of animals first captured as adults is difficult or impossible to determine accurately, only studies lasting for a decade or more can document age-specific survival and reproduction of senescent ungulates.

Very few goats on Caw Ridge survive more than twelve years: if we assume that all goats that disappeared had died, only 34% of yearling females and 5% of yearling males survived to age thirteen. But is survival to old age random? Are old individuals simply lucky? In some species of ungulates, differences in longevity can be explained by individual attributes. In fallow deer, males that have higher mating success also have higher survival (McElligott et al. 2002). In bighorn sheep and roe deer, female longevity is positively related to body mass as a young adult (Bérubé et al. 1999; Gaillard et al. 2000b), while in ibex males horn growth decreased in the two years preceding death, regardless of whether death was at six or at eleven years of age (von Hardenberg et al. 2004). Therefore, we investigated whether body mass or horn size was correlated with longevity in mountain goats.

We limited our analyses to females because the sample of males with available morphological measurements was very small. We compared age at death (including age in 2003 for sixteen surviving females aged at least ten years) with horn length at one to three years of age, with mass adjusted to July 15 at three and four years of age, and with average hind foot length at four to nine years of age. None of these correlations was significant, and some were negative.[4] Given the small sample, these results are inconclusive but provide no support for a relationship between size as a young adult and longevity.

Dispersal

It is generally believed that dispersal plays an important role in the demography of mountain goats (Hobbs et al. 1990; Stevens 1983). There are many reports of mountain goats dispersing over very long distances. In Alberta, mountain goats have been seen at Swan Hills and southeast of Edmonton, at least 300 km from the nearest goat population. A population of seven to twenty goats has become established on steep canyon walls along Pinto Creek, Alberta, isolated by about 38 km of coniferous forest from other goat populations (Penner 1988).

Caw Ridge is an island of mountain goat habitat. Although other areas with suitable habitat are visible from the ridge, goats must cross

---

BOX 9.2
Dispersal of Mountain Goats in the Olympic National Park

In the introduced population of mountain goats in Olympic National Park, Stevens (1983) documented dispersal of both sexes, almost exclusively at ages one to three. Most dispersers were males aged two or three years. Female dispersal was documented only from one high-density population at Klahhane Ridge, from which Stevens also documented the dispersal of 4.6% of goats of both sexes aged four years and older. A few adult females emigrated after having reproduced, a pattern that is unusual for mammals (Dobson 1982).

Among marked males, a minimum of 12% of yearlings, 22% of two year olds, and 18% of three year olds emigrated from Klahhane Ridge. For other populations in the park, corresponding figures were 0%, 11%, and 28%. The proportion of marked females dispersing from Klahhane Ridge declined from 10% among yearlings to 8% for two year olds and 5% for three year olds (Stevens 1983). Some goats emigrated over 100 km from their natal population. Others, particularly males, made exploratory forays of several tens of kilometers before finally emigrating. Because mountain goats were introduced to this national park and were colonizing new areas, their dispersal tendency may have been greater than in established populations. The general patterns reported by Stevens (1983), however, agree with our results.

wide tracts of boreal forest to reach them. About 14 km of unsuitable habitat must be traversed to reach Mount Hamell, with a population of approximately eighty to a hundred goats, or 13 km to reach Cutpick Hill, inhabited by perhaps a dozen goats and where at least one dispersing goat from Caw Ridge was radio tracked. Other neighboring goat populations are more distant. We surveyed Mount Hamell on foot at least six times between 1994 and 2001 but never saw a marked goat from Caw Ridge. However, in 2004, three males tagged on Caw Ridge were seen on Hamell.

The only practical way to document emigration from Caw Ridge was to conduct aerial searches for radio-collared goats that disappeared. We did not have the budget to radio-collar a sufficient number of goats or to charter aircraft to look for them, so we relied on goat or caribou censuses flown by Kirby Smith of the Alberta Fish and Wildlife Division in the area surrounding Caw Ridge. During census flights from 1993 to 2003, Kirby attempted to locate goats with radio collars that had disappeared from Caw Ridge. He would have likely picked up signals from any radio-collared goat that had moved to one of the established goat populations that were surveyed. Goats that moved to populations that were not surveyed (for example, in Jasper National Park), or to areas not known to be inhabited by mountain goats (and therefore not censused) may have gone undetected. In addition to locations during aerial surveys, we obtained a few reports of marked mountain goats seen and identified elsewhere. Nevertheless, our information on emigrant goats is incomplete, even for those equipped with radio collars (table 9.3). Known dispersers in the Olympic National Park moved an average of 41 km from their native area (Stevens 1983), therefore dispersing mountain goats from Caw Ridge could have moved over a very wide area of mountains.

Most known emigrants were aged two or three years, and males were more likely to emigrate than females (table 9.3). Two adult males may have also left, but although we picked up their radio signals away from Caw Ridge, we were unable to locate them with precision. They were never seen back in the study area. In October 1994 hunters reported seeing two marked goats walk past their camp about 8 km from Caw Ridge. The ear tag colors and numbers matched those of two females, a yearling and a two year old, that were often seen together on Caw Ridge in September but were never seen again. We did not document dispersal for females aged three years or older.

Emigration had a substantial effect on the number of young adult males in the population. The most likely time of emigration was July and August, when nineteen of thirty-two (59%) two- or three-year-old males disappeared. In contrast, for older males (which are not expected to emigrate), only 17% of twenty-three disappeared in July and August. Because few older adult goats disappeared during summer, we suspect that many of the males aged two or three years that disappeared between June and September had emigrated. A particularly dramatic emigration episode appeared to occur in summer 2000. Between August 7 and September 11, ten of eighteen males aged two or three years disappeared, but only one three year old had a radio collar. His signal was picked up from Caw Ridge and seemed to originate from a hill about 12 km away. By June 2001, however, he had returned, accompanied by another male that had left in 2000 as a two year old. The other eight marked males were not seen again. Stevens (1983) reported that most dispersing males from Klahhane Ridge left from late June to August, which agrees with our observations at Caw Ridge.

Emigration may have contributed to the high rate of disappearance of two-year-old females, but we only documented two cases of female emigration and no female immigrants during our study. Of nine two-year-old females that disappeared, five (56%) did so in July and August, while only 30% of older females disappeared during those two months.

We could monitor immigration from 1994 onward. During the first few years of the study, we could not distinguish immigrants from unmarked resident goats. After 1993, we were generally able to recognize the few remaining unmarked goats two years of age and older from individual features, especially in late summer. We could not know, however, whether unmarked two and three year olds seen in late spring were surviving residents or new arrivals. Our estimates of immigration therefore represent a minimum.

The Caw Ridge population seemed to lose more individuals through emigration than it gained through immigration: the demographic impact of dispersal was negative. We documented seven cases of immigration, all males ranging in age from two to seven years and arriving in July through September or during winter. Two other unmarked males (aged over four years) apparently spent a single day on Caw Ridge, one in late May 1996 and one in early August 1997, then left. Of the seven known immigrants, all but one were seen during at least two years and therefore were likely in the study area for at least one breeding season and could have made a genetic contribution to the population. In bighorn sheep, "breeding commutes" by rams (Hogg 2000) may have a strong impact on population genetics. There may be nonresident males arriving on Caw Ridge for the rut. The identification of paternities through molecular markers will soon shed light on this important issue.

Caw Ridge is isolated from other mountain goat habitat by wide tracts of forest, therefore the demographic impact of dispersal may be different from what one may find in populations in more continuous habitat. Dispersal had a negative but minor impact on the number of females and a moderate negative impact on the number of males. We strongly suspect that emigration of young males was a major cause of the very skewed adult sex ratio. It remains to be determined whether a similar situation exists in other mountain goat populations. If in more continuous habitats immigration rates are higher than what we documented, adult sex ratio may be less skewed.

Dispersal in young mountain goats appears to be much more common than in bighorn sheep (Jorgenson et al. 1997). Frequent dispersal of young males has also been reported for the two species of chamois (Loison et al. 1999a) and both sexes regularly disperse in Japanese serow (Ochiai and Susaki 2007). In the Les Bauges population of Alpine chamois in France it was impossible to analyze young male survival through mark-recapture monitoring, because so many males left the study area (A. Loison, personal communication). Because of its potential impact on adult sex ratio, dispersal in mountain goats has important implications for management strategies that direct the harvest at adult males.

Summary

- Cause of death was known for fewer than 20% of marked goats that disappeared. Mortality appeared to be mostly due to predation, but some goats emigrated from the study area.
- Kid survival to one year ranged from 38 to 92% and averaged 64%. It was not affected by kid sex. Mass in mid-July affected survival of female but not male kids.
- Yearling survival was greater for females (85%) than males (73.5%) and appeared slightly less variable than kid survival from year to year.
- Adult survival was greater for females than for males. For both sexes, survival was lower for two year olds than for older goats, and it showed clear evidence of senescence, for females beginning at ten years of age and for males from eight years of age. Survival of adult females was similar to that of other female ungulates of similar body size but survival of adult males appeared lower, especially for those younger than four years and older than seven years.

- Some goats, mostly aged two or three years, emigrated from Caw Ridge. Males appeared more likely than females to emigrate. At least seven males immigrated to Caw Ridge during the study, but there were no known female immigrants. It appeared that more goats emigrated from than immigrated to Caw Ridge.

Statistical Notes

1. Correlation of the proportion of kids surviving from birth to weaning and from weaning to one year for fourteen cohorts: $r = 0.28$, $P = 0.33$.

2. Comparing survival to weaning in 1989–2003 for males (91.6%, $n = 166$) and for females (84.6%, $n = 162$): $\chi^2 = 3.83$ $P = 0.05$. Survival from birth to one year: 67.1% for 146 males and 66.2% for 142 females ($\chi^2 = 0.03$, $P = 0.86$).

3. Survival from one to two years: 73.5% for sixty-eight yearling males and 84.7% for seventy-two females ($\chi^2 = 2.67$, $P = 0.10$).

4. Correlation between age at death and mass as a three year old for females: $n = 9$, $r = -0.56$, $P = 0.11$.

CHAPTER 10

Density-Dependence and the Question of Population Regulation

In 1989, there were about one hundred mountain goats on Caw Ridge. Two years later, there were 20% fewer, and the population appeared to be diminishing rapidly. By 2003, however, the population had increased to over 150 mountain goats (fig. 2.10). The population increased during all but four years between 1990 and 2003. The highest total count (152 goats, in June 2003) was 88% higher than the lowest (81, in June 1990). The number of females three years of age and older varied from thirty-two in 1991 to fifty-four in 2003, a 69% difference. In contrast, the number of males aged three years and older varied more than threefold, declining from sixteen in 1992 to only ten in 1995, then increasing to thirty-one by 2002. Can we explain some of that variability in population size? And why did the numbers of some sex–age classes vary more than others? This chapter will explore whether or not the study population showed signs of density-dependence in vital rates or in sex–age structure.

Ungulate populations limited by food availability should show density-dependence: eventually there will be more individuals than what can be sustained by the vegetation, and either some will die, some will leave, or fewer will be born (Clutton-Brock et al. 1997a). If the population does not show density-dependence, then it may be limited by something other than food, such as predation, accidents, harsh weather, or disease. We have already discussed how changes in population size over time had limited effects on the horn or body size of young goats (chapter 6) or on the age of primiparity (chapter 8). Here we will focus on population growth rate and on juvenile survival.

Arguments over population regulation have raged among ecologists for decades, partly because of misunderstandings about terminology (Sinclair 1991). Our goal here is neither to determine whether or not the Caw Ridge population was regulated, nor to assess the carrying capacity of the study area. Neither question can be answered without manipulating population size, food supply, or both. Rather, our goal is to look for evidence of density-dependence in several vital rates and in population growth rate.

In addition to its theoretical interest (Sinclair and Pech 1996), density-dependence in population dynamics plays a key role in wildlife management, because most harvesting schemes assume that mortality due to hunting will be compensated by higher recruitment. That assumption is based on the expectation that if density is lowered by harvests, recruitment will be stimulated by the lower level of intraspecific competition for resources (Caughley and Sinclair 1994). Recently, it has been pointed out that ungulate populations that are density-independent or show weak effects of density on growth rate have dynamics that are much more difficult to predict than populations showing strong density-dependence (Sæther et al. 2007). Consequently, their management is also particularly difficult.

Population Size and Forage Availability

A food-limited population should show a negative correlation between density and forage availability. For nine years between 1994 and 2003, we measured total aboveground vegetation biomass twice a year. We collected all vegetation (except for mountain avens that appear not to be eaten by goats) higher than 1 cm from twelve to fifteen 20 × 20 cm quadrats on the west end of Caw Ridge, an area heavily used by mountain goats in summer (chapter 4). We clipped plots in early June and early September, corresponding approximately to the beginning and end of the vegetation growing season. Correlations between June biomass and either the number of adult females or the total number of mountain goats in the population were not significant ($P > 0.4$) and positive, suggesting that population density had no effect on early-summer forage availability. In September, however, a comparison of forage biomass and the total number of goats in the population showed a negative trend that was nearly significant if a one-tailed test was used.[1] Vegetation biomass in early September varied among years from 22 to 71 g/m².

It would not be surprising if a few more years of data reinforced this trend: more goats consume more forage. Evidence that in years when the goat population was more numerous there was less forage biomass in late summer, however, does not necessarily imply that the population was food-limited. To suggest food limitation, we would need to show that there was not enough forage left to satisfy the nutritional needs of all individuals. If that was the case, density-dependence should be revealed through changes in population dynamics.

Density-Dependence and Population Dynamics

For all the comparisons presented here, we used the number of adult females in June as our measure of population density. As discussed in chapter 6, the number of females in June was correlated with total population size. A critical test of density-dependence in a population is a comparison of its finite growth rate λ with population size. λ is the ratio of animals at time t + 1 over the number at time t. A growing population will have a λ greater than one, a stable population a λ of one, and a declining population a λ of less than one. Population growth was not correlated with the number of adult females in June, either in the same year or in one or two years earlier.[2] Therefore we found no evidence of either a direct or a lagged relationship between population density and population growth rate.

Although population growth was independent of density, other vital rates may have been more sensitive to possible changes in intraspecific competition. Density-dependence in ungulates usually first manifests itself as a delay in age of primiparity and a decrease in juvenile survival (Gaillard et al. 1998). We already examined the relationship between population density and primiparity in chapter 8 and found no correlation. To estimate kid survival, we compared the total number of kids born one year to the number surviving to September (approximating survival to weaning) and to the number of yearlings in early June the following year. Although the proportion of kids that survived to one year varied substantially (from 38 to 92%), kid survival was independent of the number of adult females (fig. 10.1). There was also little evidence of density-dependence in kid survival to weaning, and no correlations between kid survival and the number of adult females the previous year or two years earlier.[3]

In the study of population dynamics, density is usually considered a surrogate of resource availability. Herbivore populations that are food-limited are expected to show density-dependence because, as the number of individuals increase, the amount of resources available per capita should decrease. Other factors that affect forage productivity, such as

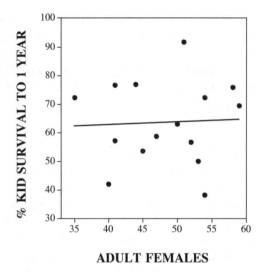

ADULT FEMALES

Figure 10.1. Survival of mountain goat kids at Caw Ridge, 1989 to 2003, compared to the number of adult females in the population in June.

changes in weather, however, should also affect population dynamics. If year-to-year variation in resource availability is important, density may not accurately reflect the amount of forage available on a per capita basis, and a composite index of forage production and intraspecific competition is more desirable. Harsh weather may also affect population dynamics either independently or in interaction with population density, as reported for other mountain ungulates such as bighorn sheep and Alpine ibex (Jacobson et al. 2004; Portier et al. 1998).

At Caw Ridge, the timing of snowmelt likely affects access to new-growth forage in spring, as suggested by the correlation between fecal crude protein in early June and body mass of young goats (chapter 6; Côté and Festa-Bianchet (2001a). Winter snowfall in Grande Cache appeared to have a negative effect on kid survival to one year, but the correlation was not significant,[4] possibly because of the small sample size or because snow accumulation in Grande Cache may not adequately reflect snow accumulation on Caw Ridge. Grande Cache is 35 km from Caw Ridge and 1000 m lower in elevation. Because negative correlations between snow cover and kid survival have been reported in other mountain goat populations (Adams and Bailey 1982; Brandborg 1955; Rideout 1974; Smith 1976), however, it is reasonable to expect that deep snow would have a detrimental effect on kid survival on Caw Ridge, particularly if years with deep snow were also years in which the snow cover lasted longer.

The effects of weather, however, are rarely independent of population density. The negative consequences of harsh weather on mountain ungulates tend to be amplified when density is high (Jacobson et al. 2004; Portier et al. 1998), presumably because poor weather and high intraspecific competition combine to reduce the amount of available resources per individual. Therefore, it would be desirable to have an index of resource availability that combines the effects of both weather and population density. For bighorn sheep, the average body mass of yearlings in early summer may be such an index (Coltman et al. 2003; Festa-Bianchet et al. 2004). In both bighorn sheep and mountain goats, yearlings are no longer dependent on milk, but have not completed body growth (chapter 6). Therefore, the mass of yearlings in early summer should be affected by the quality and availability of forage. In bighorn sheep, the mass of yearling females in early June is strongly correlated with lamb survival, ram horn growth, and age of primiparity of young ewes. Yearling mass has also been proposed as an index of resource availability for moose (Adams and Pekins 1995).

For mountain goats on Caw Ridge, there was a positive relationship between the mass of yearling males and the performance of the population. When we compared the average mid-July mass of yearling males to population growth and to kid survival, we found strong positive correlations; the population increased and kids enjoyed better survival in years when male yearlings were heavier (fig. 10.2, table 10.1) suggesting that the mass of yearling males reflects resource availability. Despite its relatively low yearly variability (range 32.9 to 38.7 kg, CV = 5.9%), average mass of yearling males explained 52% of the variability in kid survival to one year. Surprisingly, however, there were no correlations between population growth or kid survival and the mid-July mass of yearling females (table 10.1). Consequently, we must temper our conclusion that resource availability affected mountain goat population dynamics (or that changes in mass of yearling males reflect resource availability). Yearling males undergo substantially greater growth than yearling females (fig. 6.2). In early June, the average mass of yearlings is similar for both sexes, but summer mass gain is 25% faster for males than for females. Therefore, resource availability in summer may have a stronger effect on mass gain by yearling males than yearling females. That contention is supported by the finding that the summer mass of yearling males was weakly density-dependent (table 6.1), while that of yearling females was not. Because there was no correlation ($r = 0.07$) between yearly average mass of yearling males and

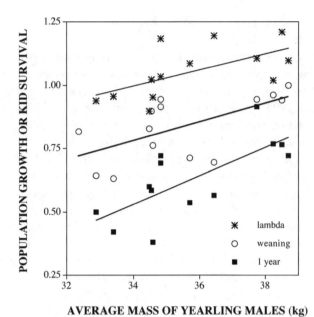

AVERAGE MASS OF YEARLING MALES (kg)

Figure 10.2. Population growth rate (λ) and kid survival to weaning and to one year compared to the average mid-July mass of yearling males in the Caw Ridge mountain goat population, 1990 to 2003.

TABLE 10.1

Pearson Correlations between Mass of Yearling Mountain Goats, Population Growth (λ), and Survival of Kids to Weaning and to One Year at Caw Ridge 1990–2003

Yearling sex	Variable	r	n	P
Male	Population growth	0.61	13	0.026
	Kid survival to weaning	0.62	14	0.017
	Kid survival to 1 year	0.72	13	0.005
Female	Population growth	−0.36	13	0.226
	Kid survival to weaning	0.18	14	0.531
	Kid survival to 1 year	0.05	13	0.875

The average mid-July mass of yearling mountain goats.

females, the factors affecting mass gain by male and female yearlings appear to be different.

In conclusion, most of our analyses suggest that changes in juvenile survival and in population growth rate were related to forage availability, but that forage availability was mostly independent of population density.

Other factors such as snow depth and the timing of snowmelt also played important roles. Although it is likely that if the Caw Ridge population continues to increase we will at some point find a negative effect of density on kid survival, fifteen years of monitoring have not provided evidence of strong density-dependence. That is a key finding with important management consequences, to be revisited in chapter 12.

Changes in Sex–Age Structure

Two aspects of the sex–age structure of the mountain goat population changed substantially during the study: the age distribution of adult females and the proportion of adult males. Recent analyses of ungulate population dynamics underline the importance of taking into consideration a population's sex–age structure to understand its dynamics (Coulson et al. 2001; Gaillard et al. 2001, 2003). That is because different sex–age classes have inherently different patterns of survival and reproduction and therefore a population's performance will necessarily vary with its sex–age composition.

Mountain goat survival varies widely with age and sex (chapters 7 and 9). Kids and yearlings have much lower survival than adults, males have lower survival than females, and older adults of both sexes show survival senescence. Changes in sex–age structure will therefore affect population growth. While the proportion of kids and yearlings in the population varied somewhat unpredictably during the study (fig. 2.10), over time both the average age of adult females and the proportion of adult males increased, two changes that should reduce population growth rate, because older females and adult males have lower survival than prime-aged females (chapter 9).

Changes in age composition can lead to spurious correlations between adult survival and population density (Festa-Bianchet et al. 2003). A simple comparison of adult female survival and the number of adult females at Caw Ridge reveals a nonsignificant negative trend.[5] Given the small sample size (thirteen years), one may be tempted to suspect that adult female survival may be density-dependent, contrary to the results of most other studies of ungulates (Gaillard et al. 2000a). Instead, this apparent trend is due to an increase in average age of adult females, as discussed by Festa-Bianchet et al. (2003). The average age of adult females increased from 5.7 years in 1992 to 7.2 years in 2003, while the proportion of females aged 10 years or more increased from 9% to 31%. Because older females have lower survival than prime-aged ones, the increase in female age led to an increase in overall female mortality, which, however, was independent of population density. An increase in average age with increasing density appears to be a common characteristic of unexploited ungulate populations (Festa-Bianchet et al. 2003). Because survival decreases with age, a greater proportion of senescent females at high than at low density would give the false impression of a density effect on female survival if age structure were ignored.

The number and proportion of males three years of age and older varied more than for any other sex–age class, from only ten in 1995 (9% of the total population and 19% of adults over three years old) to thirty-one in 2002 (21% of the population and 38% of adults over three years). Because adult males have a much lower survival (including disappearances likely due to emigration) than adult females (chapter 7), changes in adult sex ratio have substantial influences on population growth rate (Gaillard et al. 2003). Unfortunately, there are no equivalent data from other populations to determine whether these wide variations in sex ratio are typical of mountain goats. Changes in number of males appeared unpredictable, although their increase in recent years mostly paralleled a general increase in the population. The decline from sixteen males in 1993 to ten in 1994 was due to unexplained high mortality in winter. The sharp increase between 1998 and 2000, on the other hand, was due to two successive cohorts that were heavily male-biased. The population recruited nine two-year-old males (and only two two-year-old females) in 1999, and sixteen two-year-old males (and nine two-year-old females) in 2000. Apparent emigration during summer 2000 (chapter 9) reduced the number of males (including two year olds) from thirty-six in June to twenty-four by late September.

Population Regulation in Mountain Goats?

Density-dependent responses in vital rates are required to regulate populations within a range of densities (Sinclair and Pech 1996), and several studies have provided evidence of density-dependent population regulation in ungulates. Those studies generally report that when density is high, lower juvenile survival and later age of primiparity depress population growth, and when density is low, high juvenile survival and earlier primiparity accelerate growth. Over time, population density may still change considerably because of time lags and density-independent processes, but density-dependence will tend to maintain the size of the population within a certain range. Because of the frequent interaction between population density and weather, the negative effects of high density on population growth are usually more evident in years of inclement weather.

Our analyses of yearly changes in body mass (chapter 6) and in vital rates provide limited evidence that the Caw Ridge population was regulated by density and forage availability during our study. Despite an 88% increase in total population and a 69% increase in the number of adult females, there was no decline in kid survival to one year, no increase in age of primiparity, and only minor negative effects on the mass of juveniles. These results contrast with those obtained by long-term studies of marked ungulates in populations that appeared to be food-limited (Gaillard et al. 2000a). In those studies, changes in population size comparable to those observed on Caw Ridge typically led to lower juvenile survival (Clutton-Brock et al. 1997a; Gaillard et al. 1993a; Portier et al. 1998), later age of primiparity (Festa-Bianchet et al. 1995; Langvatn et al. 1996), and smaller mass or horn size of juveniles (Gaillard et al. 1996; Leblanc et al. 2001).

Why was density-dependence so weak? Over the long term, it seems highly unlikely that mountain goat numbers can be independent of forage resources. We expect that if the number of goats in our study area continues to increase, a density-dependent decrease in kid survival and possibly an increase in age of primiparity will eventually take place. For mountain goats introduced in Olympic National Park, as density increased the dispersal probability of females increased and productivity declined (Stevens 1983). In our case, however, we do not know how much more the population will have to increase before density-dependence may become detectable.

The shape of density-dependence in ungulates is rarely linear: wide changes in density can occur without any effects on per capita resource availability (Crête and Courtois 1997), and the effects of density on population dynamics may only become evident after the population reaches a threshold (Messier 1994). It appears that the goat population on Caw Ridge has not yet reached that threshold. If density-dependence played no role, the changes in population growth we observed over fifteen years were mostly the consequence of stochastic events, such as weather, changes in predation pressure, or changes in sex–age structure (Festa-Bianchet et al. 2003).

Twenty mountain goats were removed from Caw Ridge between 1986 and 1988. If there had been strong density-dependence in population dynamics, we would have expected greater population growth in the earlier years of the study, soon after the removal. Instead, the population declined slightly between 1989 and 1992 (table 2.2). Therefore, our very limited "experimental" evidence does not suggest density-dependence, or a compensatory response of the population to artificial removals. Because mountain goat hunting on Caw Ridge ended in 1969, it would have been reasonable to expect that by 1986, when the first removals took place, the population should have reached an equilibrium with its environment, even though ungulate populations may take decades to stabilize following the cessation of artificial removals (Coulson et al. 2004). The lack of compensatory response during the early years of our study suggests that the dynamics of the population may have been largely density-independent for a much longer period than the duration of our study.

One possible explanation for the lack of density-dependence in vital rates may be that the population is not food-limited. It coexists with several species of large predators, and all sex–age classes are subject to predation (chapter 9). Perhaps, predation has so far prevented the population from reaching a threshold density where food limitation may become evident (Messier 1994; Sinclair et al. 2003). Small populations of mountain ungulates may be particularly sensitive to predation by one or a few specialist predators, which may cause density-independent mortality (Gonzalez-Voyer et al. 2003; Ross et al. 1997). Predation was the main known cause of mortality on Caw Ridge (table 9.1) but did not prevent population growth. Survival of kids and adult females was not lower than that reported for similar-sized ungulates in areas with few or no predators (Gaillard et al. 1998), and therefore the impact of predation on population dynamics, although not quantified, appeared limited. Of course, the situation could suddenly change, if a resident cougar, grizzly bear, or wolf pack suddenly specialized in preying on goats, similarly to the well-documented cases of individual cougars specializing on bighorn sheep (Festa-Bianchet et al. 2006; Ross et al. 1997; Wehausen 1996). Changes in predation among different populations of mountain goats in northwestern Alberta may explain why they did not all increase following the cessation of sport hunting in 1988 (Gonzalez-Voyer et al. 2003).

Despite the absence of density-dependence, we suggest that the mountain goat population on Caw Ridge was food-limited (but not regulated) during our study. In years when fecal crude protein in early June was high, kids were heavier by midsummer. More importantly, kid survival and population growth rate were both correlated with the mass of yearling males, suggesting that in years when forage resources were fewer or of poor quality, yearlings were small, kids were less likely to survive, and the population did not grow as much as in years with more high-quality forage. Therefore, it appears that forage availability or

BOX 10.1
Can Mountain Goats Be Part of a Predator–Prey Equilibrium?

Studies of ungulates coexisting with large predators have explored predator–prey dynamics by testing theories of density-dependence in predation risk or in the numerical and functional responses of predators. Quite reasonably, those studies have mostly tested theories suggesting that as prey availability changes, so does either the proportion of that prey in the predators' diet, or the population density of predators (Linnell et al. 1995; Messier 1991, 1994; Sinclair and Pech 1996).

The dynamics of ungulate–predator systems are complicated by time lags in numerical responses of both predator and prey, the effects of alternative prey, the effects of weather, and the interactions of multiple predator species (often including human harvests), making these systems extremely difficult to understand and almost impossible to predict (Bergerud and Ballard 1988; Boutin 1992; Jedrzejewski et al. 2000; Kunkel and Pletscher 1999; Messier and Joly 2000; Post et al. 1999b; Skogland 1991; Van Ballenberghe and Ballard 1994). The predator–prey ecological relationships investigated by most of these studies, however, are likely very different from those experienced by most populations of mountain goats, and possibly by most populations of mountain ungulates in North America. That is because very few populations of mountain ungulates are sufficiently large to either sustain or be a major food source for a population of predators.

If the Caw Ridge mountain goats disappeared, it is unlikely that the density of wolves, grizzly bears, or cougars in the area would decrease, although single individuals could be affected. That is because mountain goats likely provide a small proportion of the food obtained by predators. The small number of goats (compared, for example, to moose, deer, elk, or beavers) implies that, while individual predators may benefit from developing the specialized behaviors necessary for successful predation on goats, no population of large predators can be dependent on mountain goats. Mountain goats have very different behavior and use very different habitats compared to the cervids that sustain most predator populations in the boreal forest, and would therefore require a very different hunting technique. Similarly to bighorn sheep (Festa-Bianchet 1988d, 1991) and many other mountain ungulates, mountain goat behavior suggests past selection to avoid coursing predators such as wolves. They are gregarious, particularly in open habitat, remain close to escape terrain, and avoid predation mostly by being alert and running to escape terrain at the sight of danger. Their traditional area-use patterns, consistent use of well-worn trails, and lack of preferred association with relatives (chapter 5) suggest a need to be constantly aware of where they are and what route to use in case of a predator attack, and that they find safety in numbers.

BOX 10.1
Continued

Goats share these characteristics with bighorn sheep (Festa-Bianchet 1991) and therefore, like bighorn sheep, they may be highly susceptible to predation by cougars, which use ambush to first stalk and then rush their prey, with devastating consequences for some populations (Ernest et al. 2002; Festa-Bianchet et al. 2006; Ross et al. 1997; Wehausen 1996). Although there are numerous reports of individual cougars specializing on bighorn sheep, we know of none for mountain goats. Possibly, goats are a particularly dangerous prey to subdue because of their sharp horns and aggressiveness: we twice saw goats successfully repel attempted predation by wolves. Cougars, however, regularly kill adult male elk and yearling moose (Iriarte et al. 1990; Ross et al. 1995; Ross and Jalkotzy 1996) and should not be intimidated by mountain goats. We expect that cougar predation on mountain goats would also involve an individual specialist that concentrated on this prey species. The occurrence of an individual specialist could have a very serious negative effect on mountain goat population dynamics, but it may be unpredictable and unrelated to either prey or predator density.

quality affected population growth, which was mostly independent of density. Similarly to most other long-term studies of marked ungulates (Gaillard et al. 2000a), our results suggest that only kids and possibly yearlings were affected by year-to-year changes in resource availability.

Kid production by adult females varied substantially from year to year. The proportion of females aged six to thirteen years that produced kids averaged 78%, but ranged from 57% in 2001 to 97% in 2000. We cannot currently identify the causes of variation in fecundity by adult females. Kid production appeared to be independent of weather, population density, or resource availability as measured by the mass of yearling males (table 10.2). An investigation of what causes variability in kid production by prime-aged females could shed light on the dynamics of mountain goat populations.

Changes in mountain goat numbers from year to year were due to a complex array of changes in productivity, survival, age structure, predation, and forage availability. Mountain goats on Caw Ridge showed little response to changes in density, possibly because density-independent processes affected food availability and because many variables affecting changes in population growth were independent of population density.

TABLE 10.2

Pearson Correlations between the Proportion of Female Mountain Goats Aged 6 to 13 Years That Produced Kids and the Number of Adult Females, the Average Mass of Yearling Males the Previous Year, and Total Snowfall, 1990–2003

Correlated variable	r	P
Number of adult females	0.13	0.66
Mass of yearling males	0.16	0.58
Winter snowfall	0.15	0.60

The proportion of female mountain goats aged six to thirteen years that produced kids was arcsine–square root transformed, and total snowfall was recorded in Grande Cache during the preceding November–May, 1990–2003. Sample size was fourteen years for all correlations. Females older than thirteen and younger than six years were excluded to avoid the effects of age-related differences in fecundity (chapter 7).

Although it is likely that over the very long term goat populations are maintained within a wide range by changes in food availability, our fifteen years of research on Caw Ridge provide little support for a consumptive management strategy based on the assumption of density-dependence or compensatory mortality. Weak density-dependence appears to characterize native mountain goat populations. An analysis of aerial counts of other mountain goat populations revealed that the dynamics of eight of twelve populations in west-central Alberta were similar to those observed on Caw Ridge (Hamel et al. 2006).

Summary

- We found very little evidence of density-dependence in population dynamics. Population growth, kid production, and kid survival were independent of the number of adult females in the population.
- Population growth and kid survival to one year were correlated with the average mass of yearling males in mid-July, suggesting that yearly changes in resource availability affected the population dynamics of mountain goats. Average mass of yearling females, however, was not correlated with either population growth or kid survival.
- As the population increased in size, the average age of adult females also increased.

- The number of adult males and the proportion of the population made up of adult males varied widely over time, mostly because of stochastic events.
- Predation played a limited role on population dynamics. Predation on small, isolated populations of mountain ungulates could vary with the behavior of individual predators in a density-independent fashion, and therefore may be highly unpredictable.

Statistical Notes

1. Correlation between forage biomass in September and total number of goats, one-tailed test: $r = -0.54$, $n = 9$, $P = 0.07$.

2. Correlation of population growth (λ) and number of adult females in June in the same year ($r = -0.12$, $n = 13$, $P = 0.69$), one year earlier ($r = 0.12$, $n = 13$, $P = 0.69$), or two years earlier ($r = -0.05$, $n = 12$, $P = 0.88$).

3. Correlation of proportion of kids surviving to weaning and number of adult females in June: $r = -0.19$, $P = 0.49$, $n = 15$. Correlation of proportion of kids surviving to one year and number of adult females ($r = 0.01$, $P = 0.9$, $n = 15$). Correlations between kid survival to weaning or to one year and the number of adult females the previous year or two years earlier: all coefficients were positive and all P-values > 0.65.

4. Correlation between November–May snowfall in Grande Cache and kid survival to one year: $r = -0.41$, $n = 13$, $P = 0.16$.

5. Correlation between adult female survival and number of adult females: $r = -0.36$, $n = 13$, $P = 0.22$.

CHAPTER 11

Female Reproductive Strategy and Ungulate Population Dynamics

The question of what drives population dynamics of large herbivores is important from both theoretical and applied viewpoints. Here we present a hypothesis linking female reproductive strategy and population dynamics in mountain goats and extend it to other ungulates with similar age-specific schedules of reproduction and survival. To introduce our hypothesis, we will compare the ecology of female mountain goats to that of other ungulates that have been the subject of long-term studies of marked individuals. Some aspects of mountain goat ecology are very similar to those of other ungulates: female survival is much higher and less variable than juvenile survival, the fitness costs of reproduction are mostly restricted to a decrease in future reproductive success and very rarely affect maternal survival, and lifetime reproductive success is mostly determined by longevity. We suggest that a conservative reproductive strategy, selected to maximize maternal survival, may explain why both stochastic and density-dependent mortality affects juveniles much more than adult females, and explore the implications of this hypothesis. We conclude by presenting our views on the potential impacts of hunting and predation on the evolution of female life-history strategy, and suggest that the evolutionary implications of alternative management strategies of ungulate harvest could be as important as their demographic consequences. The evolutionary implications of harvests are, therefore, an important consideration that should not be ignored by wildlife managers.

The Female's Reproductive Strategy:
Maximize Her Own Survival

Mountain goat females almost never produced twins during our study; therefore, litter size played a limited role in individual differences in lifetime reproduction. To increase their lifetime reproductive success, females could have lived as long as possible, reproduced as early as possible, reproduced every year, and maximized kid survival. Staying alive for as long as possible was the most important component of female reproductive strategy. Over three quarters of the variability in the number of kids that survived to one year was explained by differences in maternal longevity (table 7.3). In mountain goats, female longevity apparently explains a greater proportion of the variability in lifetime reproductive success than in other ungulate species such as bighorn sheep (Bérubé et al. 1999) or red deer (Kruuk et al. 1999b).

The number of surviving offspring over a female's lifetime is not necessarily an accurate measure of fitness, which may be influenced by the timing of reproduction, by changes in population size, and by differences in future offspring survival and reproductive success (Brodie and Janzen 1996; Coulson et al. 2006). In particular, offspring born early in a female's life, all else being equal, are typically of greater fitness value than those born later, except in declining populations (Hutchings 1993; Käär and Jokela 1998). For most field studies, however, the number of surviving offspring produced over a female's lifetime remains an adequate practical approximation of fitness. For mountain goat females, limited to a single offspring per year and with moderate variability in kid survival, the key to high lifetime reproductive success was longevity.

To reproduce, females had to survive several years before primiparity (fig. 7.1). Many females died without reproducing (chapters 7 and 9). Once they began to reproduce, females appeared to allocate more resources to their own maintenance than to reproduction, because with the exception of primiparous females, we found no short-term costs of reproduction (chapter 8). Although individual differences undoubtedly decreased our ability to measure the fitness costs of reproduction (Partridge 1992), similar analyses of observational data have shown reproductive costs in red deer and bighorn sheep (Clutton-Brock et al. 1983; Festa-Bianchet et al. 1998). The failure to detect fitness costs of reproduction in mountain goats may indicate a lower reproductive effort than in other ungulates, as suggested in chapter 8. After primiparity, mountain goats were more likely than females of many other ungulates to not reproduce in some years (chapter 7): on average they missed a reproductive opportunity every 4.5 years. Among other ungulates, the fecundity of prime-aged females was higher than that of mountain goats in at least 69% of fifty-eight studies reviewed by Gaillard et al. (2000a). The actual figure may be higher, as many studies underestimated fecundity because they relied on age ratios or did not monitor individuals as closely as we did during the parturition season. In years when they did not reproduce, females were more likely to care for their yearling (chapter 5), but we found no evidence that extended maternal care improved yearling growth or survival (Gendreau et al. 2005). Because reproductive pauses did not seem to benefit previous offspring, we interpret the low adult fecundity as further evidence of low reproductive effort.

Why did female mountain goats apparently restrain reproductive effort? Possibly, their reproductive strategy evolved under harsher environmental conditions than those prevalent during our study. If resources were very scarce and unpredictable, even a small amount of maternal care may affect maternal survival. Reproductive strategies may be adapted to long-term average conditions rather than responding to year-to-year variation in resource availability (Clutton-Brock et al. 1996). That interpretation is questionable, however, given the high reproductive rates exhibited by mountain goats released in unoccupied habitat, including the ability to twin and to conceive as yearlings (Bailey 1991; Houston and Stevens 1988; Swenson 1985; Williams 1999). Bailey (1991) assessed the costs of reproduction in two populations introduced into unoccupied habitat. His results were similar to ours: the only measurable cost of reproduction was that primiparity at age three reduced reproductive success at age four. Consequently, the high reproductive rates of many introduced mountain goat populations do not seem to lead to short-term fitness costs. Instead, goats simply take advantage of favorable environmental conditions by increasing reproductive output. Because of the abundant resources available in unoccupied habitat, mountain goat females in introduced populations may be able to sustain a greater reproductive rate without short-term fitness costs. Female reproductive strategy in introduced populations is likely the same as what we found at Caw Ridge, with a strong selective pressure to ensure maternal survival.

The conservative reproductive strategy of female mountain goats is likely explained by the much higher survival of reproductive females than of kids, combined with the long period of time between birth and first reproduction. On average, mature females have a much greater residual reproductive potential than kids. A female kid would have only a 45% probability to reach four years of age (when less than 50% of females reproduce), while a male's chance of surviving locally to age four (when he may have a chance to reproduce) is only 28%. About 96% of presenescent adult females, on the other hand, survive to the following year, and even for older females the chance of survival is usually higher than 75% (fig. 9.2). Of course, this reasoning assumes that kid survival is mostly independent of maternal care, and a key consideration here is how the amount of maternal care may affect a kid's survival chances. We could not quantify that effect directly because we had no reliable measure of maternal care. The weak relationships between kid mass or birthdate (two variables possibly affected by the amount of maternal care) and kid survival (Côté and Festa-Bianchet 2001a), however, suggest that variations in the level of maternal care may have a limited effect on kid survival.

The ability of mountain goats to twin when conditions are favorable, and to shut off twinning when conditions are difficult, suggests that they evolved in environments where resource availability may change over time. Forest fires or major avalanches adjacent to alpine areas may suddenly create new goat habitat. The apparently strong dispersal tendency of young goats of both sexes (chapter 9) may lead to opportunities to colonize new and previously unexploited areas, a situation where high reproductive output would be very advantageous.

Mountain goat life-history strategy presents a somewhat paradoxical mixture of traits typical of ungulates adapted to seral, frequently changing habitats (ability to twin, strong tendency to disperse) and of traits more commonly associated with species typical of very stable environments (late maturity, high survival, low reproductive effort). Their ability to switch from one life-history template to another may explain why they react so differently to sport harvests in introduced and in well-established populations.

Female Reproductive Strategy and Ungulate Population Dynamics

In ungulates, juvenile survival is affected by both stochastic and density-dependent processes that rarely affect female survival (Gaillard et al. 1998, 2000a; Sæther 1997). Most analyses of survival of adult female ungulates that took into account changes in age structure found no effects of either weather or population density (Festa-Bianchet et al. 2003; Gaillard et al. 1993a; Hjeljord and Histøl 1999; Jorgenson et al. 1997; Loison et al. 1999a; White and Bartmann 1998). Feral and reintroduced ungulates in Scottish isles may be an exception (Clutton-Brock and Coulson 2002), but even there the effects of weather and density on the survival of prime-aged females are much weaker than for other sex–age classes.

We suggest that ungulate females are selected to adopt a conservative reproductive strategy to ensure their survival, to the possible short-term detriment of reproductive success (Gaillard and Yoccoz 2003). In mountain goats, female longevity is the main determinant of lifetime reproductive success (fig. 7.6). Similar results have been reported for other ungulates: in two long-term studies of bighorn sheep and one of red deer, female longevity explained between 36 and 52% of the variability in individual lifetime reproductive success (Bérubé et al. 1999; Kruuk et al. 1999b). If the best strategy to achieve high reproductive success is to live as long as possible, then females should avoid any behavior that may compromise their survival. In mountain goats, both primiparity and the onset of senescence occur two years later than in bighorn sheep (Jorgenson et al. 1993a, 1997), suggesting that goats may adopt a more conservative reproductive strategy than bighorns. That suggestion is reinforced by the smaller weaning size of kids compared to bighorn lambs (fig. 8.1) and by the much greater propensity of mountain goat females to not reproduce in some years compared to bighorn ewes (Bérubé et al. 1999).

Mountain goats may be an extreme example of low maternal investment among ungulates. The slow growth rate of young goats (fig. 6.4) and the surprisingly weak relationship between kid mass and survival (chapter 9) could be adaptations to survive the juvenile stage despite a low level of maternal care. Mountain goats spread body growth over several years, most of it after weaning and therefore presumably independent of the direct effects of maternal care. Possibly, juvenile goats (kids, yearlings, and two year olds) direct more resources to metabolic reserves than to body growth, leading to slower growth but higher survival, a strategy used by white-tailed deer in harsh environments. White-tailed deer fawns on Anticosti Island (Québec), at high density in very poor habitat, were smaller but accumulated relatively more fat than fawns from nearby continental populations at low density and in good habitat (Lesage et al. 2001). Fawns likely traded body growth for reserve accumulation during their first year, but they never recovered the "lost" body growth. Adult deer on the island are about 40% smaller than those from their source population on the continent (S. Côté et al. unpublished data).

Possibly because in unhunted populations of ungulates most individuals are in the adult age class, the elasticity of adult female survival on population growth is much higher than that of other vital rates (Gaillard et al. 1998). High elasticity means that a small change in adult female survival would have a strong effect on population growth rate. Any factor

BOX 11.1
Comparing Natural and Sport Hunting Mortality of
Mountain Goats

There are two major differences between natural and sport hunting mortality of mountain goats: the sex–age distribution and the timing. Management policies and hunter preferences tend to bias the harvest toward males (chapter 12). The age distribution of sport-harvested goats of both sexes is radically different from that of goats that die of natural causes. The most obvious difference is that kids suffer high natural mortality (chapter 9), but are seldom if ever taken by hunters. Kids made up 48% of known deaths at Caw Ridge, but only 0.2% of mountain goats harvested in British Columbia in 1976–2004. None of the goats shot by sport hunters in the Yukon were kids, and in most jurisdictions the harvest of kids is illegal. Even when kids are excluded, differences between natural and hunting mortality remain. For both sexes but especially for males, hunters appear to avoid shooting yearlings and two year olds. Prime-aged females aged between four and nine years make up only 21% of natural mortality excluding kids, but from 56% to 83% of harvested females are in this age range. Males aged four to seven years (before survival senescence, see chapter 9) make up 18% of natural deaths and 47% of the harvest. The high proportion of males aged one to three years in the Caw Ridge natural mortality, however, is potentially misleading, because some emigrants may have survived elsewhere (chapter 9).

The harvest mortality pattern shown below would shorten life expectancy for both sexes compared to goats in unhunted populations. Without data on the proportion of goats harvested, however, it is impossible to accurately quantify the effect of sport hunting on expected longevity. Mountain goats are lightly hunted, and harvest is generally less than 4% of the population. Consequently, the hunting-induced reduction in adult life expectancy is likely much less than for most other ungulates subject to sport hunting (Festa-Bianchet 2003; Langvatn and Loison 1999; Nixon et al. 2001). An additional 4% yearly mortality beginning at age two would decrease a yearling female's chance of surviving to ten years of age from 56% to 39%. If the expected average longevity declines, the selective pressure for a conservative reproductive strategy should weaken.

Another important difference between hunting and natural mortality is in the timing. Most natural ungulate mortality occurs in late winter, when animals (especially juveniles and senescent adults) exhaust their fat reserves and either starve to death or become more susceptible to disease and predation (Bartmann et al. 1992; Burles and Hoefs 1984; DelGiudice et al. 1997). By remaining alive until late winter, individuals that eventually die compete with conspecifics for the scarce winter forage, possibly lowering the subsequent reproduction of survivors. In contrast, hunting seasons are usually from late August to October, when animals are in prime body condition. By removing animals before the winter, sport hunting decreases the level of intraspecific competition that survivors will face during the time of maximum resource scarcity. Although that effect is likely weak in mountain goats because of the low level of sport harvest, for other ungulates the fall hunting season may allow survivors to enjoy higher overwinter survival and greater reproductive

BOX 11.1
Continued

success over the following summer. The timing of hunting removals, therefore, is an important factor in the dynamics of hunted populations (Boyce et al. 1999; Kokko 2001).

Distribution (%) of Death by Natural Causes, Caw Ridge (1989–2004), and Hunting in British Columbia (1974–2004) and the Yukon Territory (1973–2004)

Sex	Age (years)	Caw Ridge (natural)	British Columbia (hunting)	Yukon (hunting)
Female	1–3	39.3 (33)	30.8 (2333)	7.8 (8)
	4–9	21.4 (18)	56.2 (4263)	83.3 (85)
	10+	39.3 (33)	13.0 (984)	8.8 (9)
Male	1–3	64.4 (56)	32.5 (5484)	16.8 (46)
	4–7	18.4 (16)	46.8 (7877)	55.5 (152)
	8+	17.2 (15)	20.7 (3488)	27.7 (76)

Box 11.1 Figure 1. Distribution of ages at death for 171 yearling and adult mountain goats that died of natural causes on Caw Ridge (1989–2004) and 24,429 yearling and adult mountain goats harvested by sport hunters in British Columbia (1976–2004). The last age class includes all animals 14 years and older for females and 13 years and older for males.

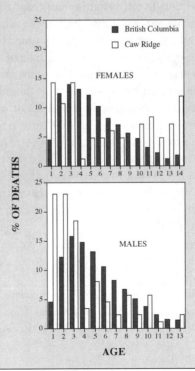

that lowered the survival of adult females would cause a rapid decline in population size and would possibly be unsustainable over the medium term. Known causes of high adult female mortality in ungulates include exotic diseases (Cransac et al. 1997; Gross et al. 2000; Jorgenson et al. 1997; Loison et al. 1996; Sinclair 1977), predation (Festa-Bianchet et al. 2006; James and Stuart-Smith 2000; Kinley and Apps 2001; Owen-Smith et al. 2005), and human harvests, both legal and illegal (Ballard et al. 2000; Festa-Bianchet 2003; Langvatn and Loison 1999; Solberg et al. 2000). Consequently, a conservative female reproductive strategy that seeks to maximize survival could explain a fundamental characteristic of ungulate population dynamics, the high and stable survival of adult females (Gaillard and Yoccoz 2003). Should that strategy be modified by changes in selective pressures, the dynamics of ungulate population would be dramatically altered. A likely source of major changes in selective pressures affecting adult female ungulates is represented by sport harvests, as considered in the next section and in box 11.1

The Potential Impacts of Harvest on Reproductive Strategy in Female Ungulates

In most unhunted populations of ungulates, females that reach the age of primiparity can expect to live through several reproductive opportunities. For mountain goats at Caw Ridge, half of the females that survived to four years reached age eleven (chapter 9) and had seven or more breeding opportunities. Similarly, more than half of the bighorn ewes that survived to two years of age at Ram Mountain or Sheep River lived for eight years or longer and had seven or more breeding opportunities (Loison et al. 1999a). The age-specific schedules of survival and reproduction of most other ungulates suggest that most primiparous females can expect six to nine reproductive opportunities. The most successful ones may reproduce twelve to fourteen times over their lifetime (Gaillard et al. 2000a). High adult survival is fundamental to our suggestion that female reproductive strategy is selected to decrease risk to maternal survival: a female should not risk dying in order to care for one offspring, because she is likely to produce several more in the future. In other words, because the residual reproductive value of an adult female is much higher than that of her offspring, she should always favor her own survival over that of her offspring.

In populations subject to sport hunting, however, ungulates face a very different age-specific probability of survival. Not surprisingly, most long-term studies of ungulate ecology are done in populations that are hunted lightly or not at all. It takes a lot of time, effort, and scarce research funds to capture and mark wild ungulates, and most researchers are not keen to mark animals that may be shot within a few months or years. That situation, however, creates two problems. One is a missed opportunity in fundamental research, because hunted populations provide an experimental test of whether life-history strategies change as predicted when age-specific survival probabilities are drastically altered. The other is that the relevance of what we know about reproductive strategies and population dynamics from studies of unhunted populations may be limited in hunted populations that have a radically different sex–age structures and sex- and age-specific survival probabilities (Festa-Bianchet 2003). Because the vast majority of ungulate populations are subject to some degree of hunting, and many are heavily harvested, our current lack of knowledge on the evolutionary impacts of hunting is a serious concern. Models of hunted populations that relied on density-dependence functions and vital rates from unhunted populations may fail because of differences in age structure, in the timing of mortality, and potentially in inherent life-history schedules that may have been selected through human harvest pressure.

From a demographic viewpoint, much of modern wildlife management is a success. There are probably more wild ungulates in both Europe and North America now than at any time over the last two centuries. For many populations of deer and elk in North America, or deer, moose, and wild boar in Europe, overabundance is a serious concern, as ungulates prevent forest regeneration, destroy habitat for other species, and affect ecosystem functions (Côté et al. 2004; Gordon et al. 2004). Many ungulate populations now exist at high density despite very high harvest rates, partly because of successful management and partly because of the seasonal nature of sport hunting, which lowers population density just before the winter, when resources are most scarce (Kokko 2001). Sport hunting, however, reduces life expectancy and increases mortality of prime-aged adults, an age class that normally has very high survival even in the presence of predators (Owen-Smith 1990; Peterson et al. 1998). A recent study comparing wolf- and hunter-killed elk in Yellowstone provides a stark demonstration of the differences in the ages selected by humans and by predators. For the same elk population, adult females killed by wolves averaged 14 years of age, those killed by hunters only 6.5 years. Young of the year accounted for 49% of wolf-killed elk, but for only 13% of elk killed by hunters (Wright et al. 2006). If a conservative female reproductive strategy is selected for by the expectation of numerous breeding opportunities, then shortened lifespan in hunted populations should select for a riskier reproductive strategy, more typical of small mammals than of ungulates (Ericsson et al. 2001).

There is little reliable information about age-specific survival of ungulates in hunted populations, but in many cases females experience a substantial reduction in life expectancy and therefore in the number of reproductive opportunities (Ballard et al. 2000; Langvatn and Loison 1999; Loison and Langvatn 1998; Nixon et al. 2001; Nygrén and Pesonen 1993). For example, fewer than 50% of yearling female white-tailed deer in a hunted population in Illinois survived to five years of age (Nixon et al. 2001). In a hunted population in Norway survival of red deer females from six months to five years was 32%, while it would have been 59% in the absence of hunting (Langvatn and Loison 1999). Moose hunting in Norway removed about 57% of females before five years of age (Solberg et al. 2000). Clearly, there are many cases where sport hunting drastically reduces the life expectancy of adult female ungulates. With much higher adult mortality than in unhunted populations, females that directed more resources to reproduction and fewer to maintenance should be favored by selection, because the fitness costs of increased reproductive effort would be discounted by the already low adult survival probability. There would not be any fitness gains from limiting reproductive effort to increase lifespan, if lifespan were cut short through hunting mortality. In addition, hunter preference for shooting females unaccompanied by juveniles may lead to a survival cost of not reproducing (Solberg et al. 2000).

In sport-hunted populations of ungulates, a conservative reproductive strategy such as that of mountain goats may not maximize fitness. Instead, a high hunting mortality rate may favor females that grow rapidly, attain primiparity at a young age, produce large litters, and provide much maternal care, including production of a large amount of milk despite the high energetic cost of lactation. Increased maternal effort should lead to higher juvenile survival, possibly confusing a compensatory demographic response to harvest with an evolutionary one. Many studies report a positive effect of artificially lowered density on the survival of juvenile ungulates (Jorgenson et al. 1993b; White and Bartmann 1998). Increased juvenile survival is likely to be primarily a demographic response to lower intraspecific competition. In ungulates, juvenile survival is typically affected by resource availability (Gaillard et al. 2000a), and sport hunting may increase resource availability by decreasing intraspecific competition. In addition, when resources are abundant, mothers can increase maternal care, including milk production, without suffering fitness costs (Festa-Bianchet et al. 1998; Festa-Bianchet and Jorgenson 1998). Over the long term, however, high hunting mortality of adult females could select for increasing reproductive investment (Ericsson et al. 2001), leading to higher juvenile survival because of greater maternal effort rather than because of greater resource availability (Festa-Bianchet 2003; Harris et al. 2002).

The demographic and evolutionary impacts of sport hunting have two important applied consequences. First, compared to unhunted populations, harvested populations have a much greater proportion of juveniles. That skew in age structure is probably more likely in species where hunters can readily distinguish between juveniles and adults, and preferentially harvest adults. That is clearly the case in mountain goats, where hunters are often prohibited from harvesting kids and appear to avoid shooting yearlings and two year olds (box 11.1). Because juveniles are much more sensitive than adults to stochastic and density-dependent mortality (fig. 11.1), hunted populations may show greater year-to-year

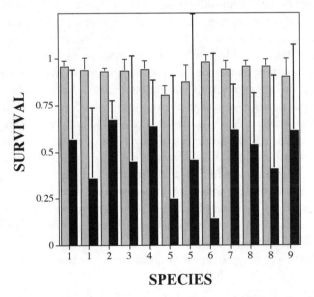

Figure 11.1. Mean and coefficient of variation of survival of juveniles (black bars) and of prime-aged adult females (gray bars) in nine species of ungulates. The "prime-age" period is from two years to the onset of survival senescence (typically eight or ten years, depending on the species). Data updated from Gaillard et al. (2000a). Two populations were monitored for some species. Species: (1) bighorn sheep; (2) caribou; (3) greater kudu; (4) mountain goat; (5) mule deer; (5) pronghorn antelope; (7) red deer; (8) roe deer; (9) Soay sheep.

variability in population growth in response to climatic fluctuations. For example, a harsh winter leading to high juvenile mortality may have little effect on an unhunted population where 10 to 15% of the animals are young of the year, but could cause a drastic decline in a hunted population where 40 to 50% of the posthunt animals are juveniles. Second, adaptation to a high hunting-induced mortality may have unpredictable consequences if harvests ceased or if some other ecologically important variable was modified, as may occur with climate change or following reintroduction or natural recolonization by large predators (Apollonio et al. 2003; Berger et al. 2003).

One such change may be under way as the number of hunters is decreasing worldwide and especially in North America (Enck et al. 2000), although so far there is little evidence that ungulate harvests have decreased because of fewer hunters. An apparently suboptimal reproductive strategy in unhunted red deer females, including substantial fitness costs of reproduction, has been attributed to the selective effects of earlier culls (Benton et al. 1995). That population is among the few where a survival cost of female reproduction has been reported (Clutton-Brock et al. 1983), and the high level of reproductive investment may in part be due to earlier artificial selection on life-history strategy through hunting of adult females. The evolutionary consequences of sport harvests are largely unknown and are likely complicated by the accompanying changes in sex–age structure (Coulson et al. 2004). Because those effects could be substantial, however, they are worth considering (Coltman et al. 2003; Festa-Bianchet 2003; Harris et al. 2002). Clearly, long-term monitoring of individuals in hunted populations could be very rewarding both for evolutionary ecology and for wildlife conservation. One interesting question to examine is whether the elimination or near-elimination of senescent adults may to some extent compensate for the greater abundance of juveniles in terms of buffering the population against weather-induced variability, given that older adults appear more sensitive to weather than prime-aged ones (Gaillard et al. 2000a; Jacobson et al. 2004). The younger age structure of hunted populations would lead to a high natural survival of adult females (Festa-Bianchet et al. 2003).

One way to decrease both the ecological and the evolutionary impacts of sport hunting would be to direct the harvest to those age classes that are subject to high natural mortality: juveniles, yearlings, and senescent individuals. There are very few species of ungulates where hunters would be able to easily recognize old animals, particularly old females. Young of the year and sometimes yearlings, however, are readily distinguishable in many species, including mountain goats. An ecologically illuminated management policy of directing much of the harvest to the young of the year has been implemented for many cervids in Europe (Solberg et al. 2000), but it frequently meets with resistance from hunters because of a cultural preference to harvest adults or the misguided impression that juveniles have high reproductive potential (Festa-Bianchet 2003). A management policy directing harvest to mountain goat kids would be ecologically sensible but socially very difficult to implement, as kids are small and extremely visually attractive.

Summary

- The reproductive strategy of female mountain goats appears selected to maximize maternal survival, possibly to the detriment of current reproduction.
- Mountain goats may have a more restrained reproductive strategy than most other ungulates. A low level of maternal effort may be combined with a juvenile strategy of slow, multiyear body development, favoring maintenance over rapid growth.
- In other species of ungulates, females also adopt a conservative reproductive strategy. A risk-averse female reproductive strategy may contribute to explaining why adult female survival is both higher and less variable than juvenile survival, a key characteristic of ungulate population dynamics.
- In hunted populations of ungulates, female life expectancy is reduced through harvest mortality. Consequently, females in hunted populations have fewer breeding opportunities and may be selected for a riskier reproductive strategy, possibly leading to greater juvenile production and survival.
- Wildlife managers are justifiably concerned with the demographic and habitat impacts of ungulate harvesting strategies. The evolutionary consequences of the high adult mortality rates in hunted populations should also be considered in management decisions.

CHAPTER 12

Management and Conservation
of Mountain Goats

Mountain goats are the only North American ungulates to be extirpated from large parts of their range primarily through sport hunting (Glasgow et al. 2003). They disappeared from a wide area of southern Alberta in the apparent absence of habitat loss or introduced pathogens, and suffered substantial declines in much of their historic range in the northwestern United States and southwestern Canada (Côté and Festa-Bianchet 2003). Ignorance of mountain goat ecology was largely responsible for these declines. Wildlife managers did not have the necessary information to set appropriate harvest quotas. They assumed that the population dynamics and sustainable harvest potential of mountain goats were similar to those of other species such as bighorn sheep or deer. By the time managers realized that mountain goats were much more susceptible than other ungulates to overharvest, in some areas it was too late. Today, much uncertainty remains about how mountain goats should be managed, partly because their susceptibility to harvest varies substantially among populations, and even within the same population over time (Côté et al. 2001; Smith 1988b; Williams 1999). Here we combine results from Caw Ridge with information from the literature to review current efforts to manage mountain goats for a sustainable sport-hunting yield, and provide recommendations for the conservation and management of this species.

Management of Native Mountain Goats for Sport Harvest

There is considerable interest in hunting mountain goats. In 2003 in Alberta there were 3222 applicants for seven licenses. The success rate was

BOX 12.1
Counting Mountain Goats and Monitoring Populations

Accurate and updated population monitoring is essential for the management of species that are highly sensitive to harvest. Aerial surveys, usually from helicopters, are used to monitor numbers of mountain goats in most of their range. Because Caw Ridge was surveyed by helicopter in early July during most years, we compared what was seen from the air to our complete classified counts (Gonzalez-Voyer et al. 2001). During helicopter surveys, two observers attempted to count all goats, by flying over all known goat habitat. Goats were classified as kids, yearlings, and adults. Reanalysis of the data in Gonzalez-Voyer et al. (2001), plus three more years of surveys, confirms the major conclusions of that paper. In open habitat, goats are easily seen from the air during summer, when their white color contrasts with the background. On average, 69% of the population was seen during helicopter surveys. The proportion seen during surveys was not correlated with the number of goats on Caw Ridge ($r = 0.05$, $P = 0.87$, $N = 14$ years). Therefore, survey efficiency was independent of population density.

Although aerial counts provided a reliable index of population trends, the proportion of goats seen in any one year varied from 55 to 84%, and the age classification was unreliable, likely because some small yearlings were classified as kids. In two of fourteen years, observers counted more kids than the number known to be on the ridge.

Age ratios are often recorded during ungulate counts, but their usefulness has been questioned (Festa-Bianchet 1992; McCullough 1994). For mountain goats at Caw Ridge, age ratios (such as kid:adult females, kid:total population, yearling:adult females, or yearling:total population) did not predict changes in population size, because population growth was not correlated with any of these ratios in either June or in August–September (Côté et al. 2001). Somewhat surprisingly, all correlation coefficients were negative and none was significant ($n = 13$ years, from 1990 to 2003). Age ratios appear to be of little utility for predicting population changes over the short term. For species where the sexes are almost indistinguishable during aerial surveys, classified counts are both inaccurate and imprecise and may not provide any additional information to total counts. Although many people are confident of their ability to assess age and even sex of mountain goats during aerial surveys, we are not aware of any tests of that confidence anywhere else. We suggest that tests are urgently needed, for mountain goats and for other ungulates where sex–age classification is routinely done during aerial surveys. Until it is tested, people's ability to classify ungulates by sex and age during aerial counts remains based only on faith.

Despite these shortcomings, aerial surveys are an effective population-monitoring tool over the long term, and in many remote areas they are the

BOX 12.1
Continued

only available monitoring technique. Ground surveys are not cost-efficient because of the time required to reach mountain goat habitat on foot. A combination of aerial surveys and ground counts would likely improve precision. In northwestern Alberta, when large groups of goats are encountered, the helicopter lands and classification is done from the ground using a spotting scope. An additional advantage of this technique is that it reduces the need to harass large groups of mountain goats by flying repeatedly over them in an attempt to obtain an accurate count.

Another promising monitoring technique may involve asking hunters or game wardens to report how many goats they see during hunting trips or backcountry patrols. Similarly to aerial surveys, counts by hunters or wardens will be of little use over the short term but could be very useful if collected systematically over several consecutive years while accounting for changes in search effort (Pettorelli et al. 2007) and may be a low-cost alternative to surveys in remote areas (Veitch et al. 2002).

Box 12.1 Figure 1. A comparison of the total number of mountain goats seen during helicopter surveys of Caw Ridge in early July from 1989 to 2003 and the number known to be on the ridge at the time of the survey. (Updated from Gonzalez-Voyer et al. 2001.)

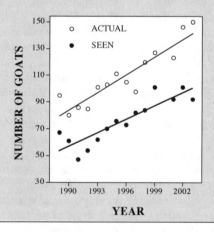

0.2 percent (%). A person diligently applying each year for forty-five years would have a 9% chance of being drawn once. In 2001 in Oregon, there were 4800 applications for four goat permits: the chances of being drawn were less than one in a thousand (Coggins 2002). Although these are extreme examples, in most areas demand for goat hunting vastly ex-

BOX 12.1
Continued

Box 12.1 Figure 2. The number of mountain goats seen during helicopter surveys of Caw Ridge in early July, 1989 to 2003, compared with the number present on the ridge by age class. Lines indicate linear regressions, which were significant for the total number of goats ($r^2 = 0.78$; $P < 0.001$) and the number of adults ($r^2 = 0.73$; $P < 0.001$), but not for the number of yearlings ($r^2 = 0.17$; $P = 0.15$) or of kids ($r^2 = 0.26$; $P = 0.06$).

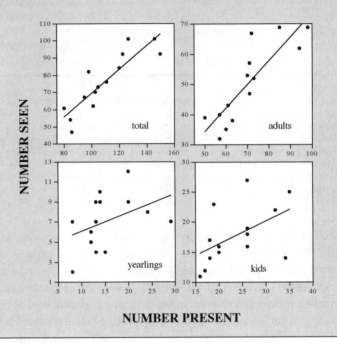

ceeds supply. Currently, mountain goat hunting permits are issued by lottery everywhere except for remote parts of British Columbia.

Most mountain goat populations outside protected areas are subject to sport harvest. Most jurisdictions with native goat populations have recently implemented policies to decrease harvest, and many managers have expressed concerns about declines. Management plans typically recommend harvest rates of 4% or less of the total population. In most jurisdictions, regulations encourage the harvest of adult males. A review of recent practices suggests that the harvest in most jurisdictions is much lower than 4% (table 12.1). Harvest rates are substantially higher in Montana than elsewhere, but the data for Montana include some

TABLE 12.1.
**Current Harvest Programs for Native Mountain Goats in
Various Jurisdictions**

Jurisdiction	Goat population estimate[a]	Yearly harvest	Harvest rate[b]	Years
Northwest Territories	698–919	4	0.5	1992–2001
Alberta	250	3	1.2	2001–2003
British Columbia	39,000–67,000	708	1.3	1999–2002
Alaska[c]	3825	61	1.6	1997–2001
Yukon	800	10	1.2	1999–2003
Washington	800	23	2.9	1999–2003
Oregon	390	3.5	0.9	2000–2001
Montana[d]	2300–3050	211	7.9	1995–1999

[a]Only areas open to goat hunting.
[b]Percent of the total goat population in areas open to hunting that is harvested each year, calculated based on the average number of goats when a range in estimates is provided. For most jurisdictions the overall harvest rate is somewhat misleading, as many goat populations in remote areas are subject to a much lower level of harvest, and some accessible populations to a much higher harvest rate.
[c]Data for Alaska refer only to Management Unit 6 that had the most long-term count records.
[d]Data for Montana include both native and introduced populations.
Note: Data were extracted from government reports and Web sites or were provided directly by wildlife managers.

introduced herds with much higher resilience to harvest (Swenson 1985; Williams 1999) and some that are heavily hunted to avoid expansion into Yellowstone National Park (Lemke 2004) where mountain goats are considered an exotic species. Province- or statewide harvest rates, or even those within a management unit, however, can be misleading, as the actual harvest pressure can vary substantially among populations. The high site fidelity of mountain goats, particularly of adult females, implies that their management must be on a herd-specific basis. Consequently, identification of distinct goat populations is essential. Otherwise, hunters may overharvest easily accessible herds and not harvest remote ones within a management unit. The rarity of female dispersal and very limited evidence for density dependence in population dynamics suggest that overharvest in one population will not increase immigration from neighboring populations.

Several aspects of the population ecology of mountain goats make them highly susceptible to hunting. Recruitment is limited by the late age of first reproduction and the high frequency of reproductive pauses in adult females (chapter 7). Although juvenile survival is apparently higher than in many other ungulates (chapter 9), the survival of yearlings and two year olds appears to be lower than in most other species. There is very little evidence of density-dependence in native populations, suggesting that hunting mortality is likely additive (Côté et al. 2001; Hebert and Smith 1986; Kuck 1977). Consequently, current management policies aimed at strongly limiting sport harvest and directing most of it to males are justified.

Unfortunately, selective male harvest meets with two problems. First, the field identification of sex in mountain goats is not easy (Smith 1988a) and requires training. Second, the number of adult males in mountain goat populations is limited by a combination of high mortality and juvenile dispersal. Between 1994 and 2003 at Caw Ridge, the average number of males aged five years and older was 10.1 (range 5–21) and the yearly recruitment of five-year-old males averaged 2.9 (range 1–6). Total population in September (the typical timing of hunting seasons) averaged 114 goats (range 92–139). A reasonable management policy to limit the effects of harvest on male age structure may be to harvest half of the yearly recruitment of males reaching five years of age. By age five males have nearly reached their asymptotic body mass (chapter 6) and presumably participate in breeding. For Caw Ridge, the largest goat population in Alberta, that policy would allow an average harvest of 1.5 goats a year, just over 1% of the population (Hamel et al. 2006). Between 1994 and 1996, the harvest of two males five years of age and older each year would have removed between 33 and 40% of all males older than five years. If completely additive, that harvest would have left only two males aged five years and older by 1996.

Harvest of mature females leads to declines in native mountain goat populations (Glasgow et al. 2003; Hamel et al. 2006; Hebert and Smith 1986; Smith 1988b). Hunter preference for large-horned individuals may partly explain why female harvests tend to be unsustainable. By selectively harvesting long-horned females, hunters remove the prime breeders from a population (Côté and Festa-Bianchet 2001d). Consequently, a challenge for both managers and hunters is to direct the harvest to males. Before receiving a license, hunters should demonstrate their ability to recognize the sex of adult goats in the field. A double-quota system should be implemented, with a quota for males and one, much smaller, for females. Hunting should be closed for one or two years in management units where the female quota is exceeded, to allow the population to recover. The double-quota system has been adopted in Alberta (Glasgow et al. 2003).

Because of their naturally high and variable mortality, juvenile ungulates can typically sustain a much higher rate of harvest than adults

(Nygrén and Pesonen 1993), and management policies directing a substantial part of the harvest to the young of the year have been implemented successfully for a number of cervids, especially in Europe (Solberg et al. 2000). Clearly, if management seeks to maximize hunting opportunities while having the least possible impact on a mountain goat population (both in terms of numbers and of age-sex structure), sport hunting should be restricted to kids, or to kids and yearlings. Most hunters would be reluctant to shoot mountain goat kids or even yearlings, because they are small and have small horns. Mountain goats certainly do not "require" sport hunting, contrary to many populations of cervids. High-density, predator-free deer and moose populations could cause economic problems and environmental degradation if sport harvest was ended, and many cause those problems despite hunting (Côté et al. 2004). Native mountain goat populations, however, do just fine if they are not hunted. There are no ecological justifications for harvesting mountain goats. Sustainable sport hunting of goats is acceptable because it provides recreational opportunities for hunters and economic opportunities for guides. If the only sustainable harvest was of kids and yearlings and hunters were not interested in shooting them, there would not be any justification for a hunting season. It is also unlikely that nonresident hunters would pay several thousand dollars to put the stuffed head of a kid on their wall. In the future, hunting ethics may evolve to the point that the experience of a difficult mountain hunt could be rewarded by the "ecologically correct" harvest of a kid or yearling. Until then, a realistic management strategy must seek a sustainable way of harvesting adults.

Our data on survival and dispersal of young males suggest that the harvest could be increased if it was directed at two-year-old males. Although two-year-old males only reach about 60% of asymptotic body size, by September their horns are over 90% as long as those of fully grown adults (chapter 6). They may therefore be an acceptable target for many sport hunters. Over the last ten years the Caw Ridge population recruited an average of seven two-year-old males each year (range 2–16), more than twice the number of five-year-olds. Although there is no information on whether hunting mortality of young males would be additive or may lead to a decrease in emigration, it is possible that fewer males would emigrate as three year olds if their numbers were reduced through harvest of two year olds. If that was to happen, at the local population level harvest mortality of young males may be partly compensatory.

Although biologically the harvest of two-year-old males is an appealing alternative, from a practical viewpoint it would be difficult to manage, because it would require identifying two-year-old males, typically within nursery groups. In Canada, such a requirement may well be legally shaky, as judges may decide that it places an unreasonable onus on hunters. To harvest two-year-old males, hunters would have to look for young males within nursery groups, which may lead to a high incidental harvest of adult females. It is questionable whether such an age restriction could be enforceable, and it may be very difficult to convince hunters to shoot two year olds if older and larger males were available within the hunting area. Wildlife management rules, to be successful, have to be both enforceable and acceptable to a majority of hunters.

Male-biased harvest of mountain goats may have a number of ecological and evolutionary consequences about which we currently know very little. Those consequences may be particularly important if the sex–age structure we recorded is representative of most mountain goat populations. At Caw Ridge, there were so few mature males that any sport-hunting removal would have strongly affected the operational sex ratio. Lack of knowledge further argues in favor of a very conservative harvest strategy. If too many males are removed, some females may fail to conceive or may conceive late, as recently suggested for moose and reindeer (Holand et al. 2003; Sæther et al. 2003). The risk of late conception may be substantial in small populations where females are solitary or in small groups. If many mature males were removed through hunting, the opportunities for mate choice by females would decrease. There are currently no data on any effects of mate choice by female mountain goats, so it is impossible to assess what role, if any, it plays in the population genetics and population dynamics of this species. If survival is partly correlated with genetic quality, however, mature males could be of above-average genetic quality and may therefore be preferred mates. An extreme level of removal of mature males may have unwanted consequences, but a moderate removal would still allow female choice among surviving males.

Hunter preference for large-horned bighorn rams, combined with regulations that protect small-horned rams, leads to artificial selection favoring rams with small horns (Coltman et al. 2003) and possibly removing good genes from hunted populations (Coltman et al. 2005). Artificial selection through trophy hunting is unlikely to be a major concern for mountain goats, for two reasons. First, there is little evidence that horn size affects dominance relationships among adult males. Second, the limited variance in male horn size (chapter 6) should reduce the potential for hunter selectivity: it is unlikely that many hunters will be able to distinguish differences in horn length of the order of 1 to 2 cm, unless several males are in the same group. In addition, any hunter selectivity will likely focus on horn rather than body size, and body size may be a more important determinant of male reproductive success than horn size. It is unclear, however, how the removal of mature males may affect the social hierarchy among survivors. If the most dominant males are removed, there may be more escalated fights because most surviving males would be of similar size. That could lead to an unstable dominance hierarchy, as reported for pronghorn antelope following high mortality of mature males (Byers and Kitchen 1988). Escalated fights and unstable dominance relationships may lead to a greater frequency of wounding during the rut and possibly increase male mortality. Finally, male harvest will further bias the already skewed adult sex ratio and therefore reduce effective population size, possibly increasing the risk of inbreeding. Research on the reproductive behavior and mating success of male mountain goats is urgently required to assess the possible impact of male-biased harvests on population ecology, genetics, and behavior.

Metapopulation Dynamics and Goat Management

Since the mid-1970s, biologists from the Alberta Fish and Wildlife Division have conducted regular helicopter surveys of several mountain goat populations in northwestern Alberta. Most hunted populations declined until hunting was closed in 1987, then recovered slightly between 1988 and 2000 (Gonzalez-Voyer et al. 2003). There was little synchrony in dynamics among populations, suggesting that large-scale environmental factors shared by all populations, such as weather, likely played a limited role. Instead, differences in numerical trends were likely due to local differences in habitat quality, predation pressure, dispersal, or population age structure. We examined changes in total population size, as estimated by helicopter surveys, for twelve populations, including Caw Ridge. None of these populations were hunted after 1987. Overall, in the last four years of surveys (2000–2003) about 27% more goats were seen than during the first four years (1988–1991) (fig.12.1).

Although the overall increase may suggest a population recovery and justify reopening a limited hunting season, a closer look at the survey data reveals substantial differences in numerical trends among populations (fig. 12.1). Of the four populations hunted before 1987, two increased, one was stable, and the other first increased then declined. Of unhunted populations, five (including Caw Ridge) increased, two declined, and one was stable. These data suggest that mountain goat populations must be managed on a herd-specific basis, especially given that differences in accessibility lead to major differences in vulnerability to

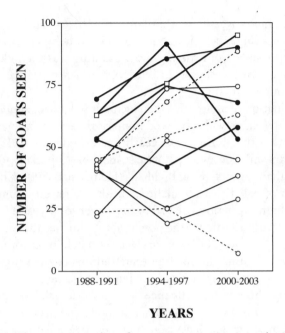

Figure 12.1. The average number of mountain goats seen in twelve populations in northwestern Alberta (all within 100 km of Caw Ridge) during helicopter surveys conducted in July over three four-year periods. Filled circles indicate populations that were hunted before 1988. The open square symbols indicate counts for Caw Ridge. All other populations were not hunted since at least 1970. At least two surveys were conducted during each four-year period for each population.

hunting (Hamel et al. 2006). Of the twelve herds on figure 12.1, two are visible from a paved road, and Caw Ridge is accessible by all-terrain vehicle. Other herds, deep inside the Willmore Wilderness, can only be reached after one or more days of hiking or horseriding. Wisely, none of the easily accessible herds have been open to hunting since 1970.

An anecdote from southern Alberta illustrates the difficulties of distributing sport harvest among herds. In the Kananaskis area, one population of mountain goats had increased to over one hundred by the mid-1980s. Based upon hunting statistics from the Willmore Wilderness, managers issued six permits, expecting a success rate of about 40% and a harvest of three goats, less than 3% of the population. Most of the target herd used areas accessible with a day's hike, but one slope used by about twelve to fifteen goats (possibly a separate subpopulation) was visible from a major highway. All six hunters that received a permit spotted a goat from the highway, and each shot one from that small subgroup. The following year a single permit was issued and another goat was shot from the subgroup visible from the highway. Hunting was then closed. The best way to manage easily accessible populations of native mountain goats is to not hunt them.

An examination of alternative management scenarios combining vital rates from Caw Ridge with observed numerical trends from nearby populations suggests that nonselective yearly harvest rates above 1% of goats aged two years and older would not be sustainable, especially for populations of less than seventy-five individuals (Hamel et al. 2006). Those simulations also revealed that populations of fifty or fewer mountain goats would have high extinction risk (18 to 82% over forty years) even in the absence of harvest. Elasticity (the potential to influence the rate of population changes [Caswell 2001]) in survival was much lower for males than for females, illustrating the small potential influence of changes in male survival on population dynamics. For females, elasticity increased with age, outlining the greater influence of mature females on population growth rate. In other words, the same proportional decrease in survival of mature females would have a greater negative effect on population growth rate than if it affected very young females. Thus we recommend a 1% harvest rate for native mountain goat populations, provided that all or most of the harvest is made up of adult males (Côté and Festa-Bianchet 2003) and that only populations of seventy-five or more goats are hunted (Hamel et al. 2006). That very low rate of sustainable harvest provides little margin for error. Because most populations are small, overharvest of just one or two goats will more than double the proportion removed. Because of the imprecision of aerial surveys (Gonzalez-Voyer et al. 2001), however, managers could only detect major declines. Slow but steady declines may not be detected until after they have caused substantial reduction in population size. Several years of protection will then be required for population recovery.

Management of Introduced Mountain Goats for Sport Harvest

Some populations of mountain goats introduced outside their native range harvested at rates of 7 to 10% or more have either stabilized or continued to increase (Adams and Bailey 1982; Houston and Stevens 1988; Swenson 1985; Williams 1999). Those harvest rates would lead to rapid declines in native populations. Williams (1999) reported compensatory reproduction in an introduced and intensively hunted population in Montana and recommended a harvest rate over 15% for introduced populations. However, his calculations for compensatory reproduction included mathematical errors (Côté et al. 2001). Other studies suggest that a harvest rate of less than 7% is more realistic for introduced populations if the goal is sustainable hunting.

Several introduced populations of mountain goats have shown the irruptive patterns typical of other ungulates introduced to new habitat (Caughley 1970). Those populations show a young age of primiparity (typically three years and sometimes even two years) (Bailey 1991) and a high frequency of twinning (Houston and Stevens 1988), leading to high recruitment and population growth. Introduced populations often have access to highly productive habitats with abundant forage and few or no competitors, no or few predators on adults, and possibly forage species not previously exposed to ungulate herbivory, that may lack defenses such as secondary toxic compounds (Bryant et al. 1991; Robbins et al. 1991). Most introduced populations are south of their historic range, where the climate is generally milder than in our study area. Introduced mountain goats can rapidly increase in the absence of intense harvesting pressure (Houston and Stevens 1988).

The population originating from a dozen mountain goats released in the Olympic Mountains of Washington in the 1920s has led to a controversy, with different people calling for either the conservation of goats or their eradication (Houston 1995; Hutchings 1995; Lyman 1994, 1999; Scheffer 1993). Those favoring eradication, including the U.S. National Park Service, argue that grazing by mountain goats has a negative impact on the fragile alpine vegetation. Those against eradication suggest that the effects of goats on the vegetation are minimal. A control program greatly reduced the population until 1989, but it has since been discontinued. Concerns have also been raised about the possible ecological impacts of mountain goats in Yellowstone National Park. Goats are not native to Yellowstone, but were introduced into several adjacent mountain ranges in Montana, from where they have immigrated into the park (Lemke 2004).

Because of their potential impacts on native species, exotic populations of mountain goats should be subjected to a high level of sport harvest. In some cases, eradication or major depopulation of introduced goats may be required to protect biodiversity, especially in national parks where sport hunting is not an acceptable management tool (Houston and Schreiner 1995). The evidence that hunted exotic populations of mountain goats have a negative impact on biodiversity, however, is mostly restricted to the coastal ecosystem of Olympic National Park.

Most exotic populations have not reached such high densities that they could impact plant and other animal species or affect ecosystem functions, and can be kept at low density through sport harvest. Their continued management for sport hunting does not appear incompatible with conservation.

Conservation of Mountain Goats

With careful monitoring to prevent overharvest, sport hunting of mountain goats can be sustainable and is an acceptable activity. We do not advocate sport hunting, however, for all populations of mountain goats. Most native populations coexist with large predators and have a slow rate of growth. They did quite well before modern wildlife management, and are likely to continue to do well if their habitat is protected. Hunting can be an important part of a conservation strategy, but habitat protection and the preservation of the mountain goat's ability to evolve and to persist over the very long term must take precedence over consumptive use.

Conservation of mountain goats faces two related challenges: their habitat must be protected, and they must continue to use available habitat. Because mountain goats are highly intolerant of human activities, their conservation requires protection from human intrusions. Otherwise, goats may not use available habitat. Mountain goats are known to be very susceptible to helicopter harassment (Côté 1996), but other forms of motorized access can also affect their behavior, increasing movement rate, decreasing foraging time, and in extreme cases leading to habitat abandonment. Intense recreational use of their habitat, especially with motorized access, is incompatible with mountain goat conservation. In chapter 4 we argued that abandonment of some areas may cause population declines in two ways: by decreasing the total amount of forage available and by increasing predation risk, if goats were forced to use small areas and be more predictably located by predators. The expansion of industrial activities and motorized recreation (mostly helicopters, but also four-wheel-drive and all-terrain vehicles, snowmobiles, and in a few cases boats and ski resorts) into mountain goat habitat is a major concern for the conservation of this species, even if it does not lead directly to habitat destruction.

While consumptive management of mountain goats requires a herd-specific approach, their conservation also requires the identification of seasonal ranges, of the travel routes connecting those ranges, and of the likely dispersal routes between populations. The frequent dispersal of young goats (chapter 9) suggests that mountain goat conservation must include a metapopulation approach (Hanski 1999) to maintain opportunities to disperse among populations.

Currently, many populations of mountain goats are remote and therefore under little immediate threat. Industrial activities, particularly mining and exploration for oil and minerals, increase motorized access through the development of a network of access roads. The expansions of these activities and the ever-growing network of road in former wilderness present increasing threats to mountain goats. The construction of skiing facilities and the increasing popularity of heli-skiing and other helicopter-based activities are also a menace to the conservation of mountain goats.

The potential impacts of global warming on mountain goats are not understood but are likely to be severe. Global warming is expected to have a particularly strong effect on the northern and alpine areas where mountain goats are found (Hughes 2000). The continuing trend toward warmer climate may make some areas unsuitable as habitat for mountain goats. Some biologists predict that global warming may soon overtake habitat destruction as the leading cause of extinctions, particularly in mountainous environments (Thomas et al. 2004). Continued monitoring of mountain goats on Caw Ridge will allow us to assess the potential effects of climate change.

Summary

- Mountain goats are the only North American ungulate that suffered extirpations and major population declines primarily through poorly regulated sport hunting.
- Aerial censuses can monitor mountain goat populations over the long term by detecting major population trends. Over the short term, however, aerial surveys are not very useful because of wide variation in the proportion of goats seen. Age and sex classifications obtained during helicopter surveys are unreliable. Age ratios provide little insight over short-term changes in population size of mountain goats.
- Native mountain goat populations are sensitive to overharvest if adult females are shot. They have a low natural recruitment rate and show little evidence of density-dependence or of compensatory responses to hunting. Hunting mortality appears additive.
- Current management of native populations of mountain goats is conservative. Most jurisdictions have harvest objectives of 4% or less of the total population, and in most cases the harvest is less

than 2%. All jurisdictions encourage the harvest of adult males. Regulated sport hunting is not a threat to the persistence of mountain goat populations.

- Because of the strong site fidelity of mountain goats, a major challenge confronting managers is to spatially distribute hunter harvest among herds with major differences in accessibility. Accessible herds are easily overharvested and are best managed for nonconsumptive use.
- The conservation of mountain goats rests on the protection of their habitat and on the prevention of harassment, particularly from helicopters and motorized vehicles. Like other alpine species, mountain goats may soon be negatively affected by global warming.

CHAPTER 13

Long-Term Monitoring of Marked Individuals and Advances in Ecology and Conservation

We have emphasized two aspects of our study of mountain goats: its relevance for the understanding of fundamental ecological principles and its applications to the conservation of ungulates in mountain ecosystems. The greatest asset of our research was its reliance on monitoring marked individuals over their lifetime. To conclude, we will first reflect on the contributions of long-term studies of marked individuals to the understanding of basic ecological processes, then consider how studies addressing fundamental ecological and evolutionary questions can be useful to manage wildlife and conserve biodiversity. Those considerations will be followed by an overview of the main future research questions at Caw Ridge, addressing gaps in our knowledge of ungulate ecology. We will conclude with a consideration of the future of Caw Ridge and of its rich alpine biodiversity, including our study population of mountain goats.

Long-Term Studies of Marked Individuals and Their Contribution to Ecology

Few of our results could have been obtained had we not monitored marked individuals over several years, or if we had not marked nearly all mountain goats in our study population. The ecology of large mammals can best be understood with long-term studies, because large mammals are long-lived, they distribute their reproductive success over many years, and adults are resilient to short-term changes in their environment, other than the drastic alterations caused by humans. For studies of population dynamics, each year produces a single data point.

BOX 13.1
The Unreliability of Short-Term Studies of Population Dynamics

The unreliability of short-term studies of population dynamics can be illustrated by comparing the correlations between the number of adult females in the Caw Ridge population and kid survival over the entire study with those calculated for any consecutive five- or ten-year periods. Over the entire study, the number of adult females was not correlated with either kid survival to weaning (15 years, $r = -0.19$, $P = 0.5$) or to one year (14 years, $r = -0.04$, $P = 0.9$). Over different ten-year periods, however, the correlation between number of adult females and survival to weaning ranged from -0.46 to -0.01, while that between number of females and survival to one year varied from -0.35 to -0.08. None of these correlations was significant. When five-year intervals were considered, the range in correlation coefficients was much wider, from -0.83 to 0.36 for survival to weaning and from -0.88 to 0.30 for survival to one year. With five data points, a correlation coefficient of -0.88 is significant ($P = 0.049$). All ten-year periods and most five-year periods suggested a more strongly negative correlation between density and kid survival to one year than that obtained when all years were considered. Even fifteen years of data may be insufficient to reliably assess density-dependence in juvenile survival, as shown by the longer time series available for bighorn sheep at Ram Mountain. In that population, lamb survival decreased with increasing numbers of ewes (Portier et al. 1998), as was evident for most comparisons using twenty years of data or more. Over twenty-seven years, the correlation coefficient was -0.53. When only fifteen years of consecutive data were examined, however, correlations ranged from -0.84 to 0.02, and five of thirteen coefficients were not significant.

The results presented in the figures in this box are predicted by statistical theory (Sokal and Rohlf 1981): as sample size decreases, errors in sampling increase random departures from the "true" underlying trend. Because they are based on empirical measurements, however, these comparisons illustrate the low reliability of short-term studies of population dynamics, and justify a healthy skepticism toward low sample sizes. They also illustrate the problem of publication bias (Palmer 1999, 2000). It is not unreasonable to suspect that a study claiming that the correlation between lamb survival and population density was -0.84 would attract more interest (and perhaps be published in a better journal) than one reporting a correlation of 0.02, even if the two studies had the same duration and sample size. Studies that rely on natural variation may not encompass a sufficiently wide range of circumstances to draw clear conclusions about questions such as density dependence. Finally, some short-term trends may reflect real ecological phenomena, and long-term studies may be able to determine why density-dependent effect can vary over time, for reasons such as changes in age structure or in population-limiting factors.

BOX 13.1
Continued

Box 13.1 Figure 1. Correlation coefficients (r) between the number of adult female mountain goats at Caw Ridge and the survival of kids to weaning and to one year of age, according to the number of years of data used for the correlation.

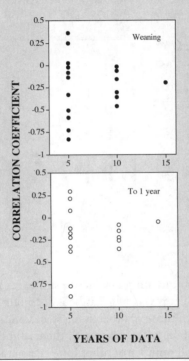

YEARS OF DATA

Particularly when relying on natural variation, one cannot analyze the effects of weather or density on population dynamics before accumulating many years of monitoring. Several years of data are required just to establish what is "normal" in terms of yearly variability in population parameters and environmental effects. Short-term studies often report unusual events, but those events may not seem quite as unusual if a longer-term perspective is available (Weatherhead 1986). To understand population dynamics and reproductive strategies, yearly variability in various parameters is as important as their mean value, because temporal variability is important in assessing the risk of extinction and the fitness benefits of alternative strategies (Beissinger and Westphal 1998; Caughley 1994).

BOX 13.1
Continued

Box 13.1 Figure 2. Comparison of the correlation coefficients (*r*) between the number of adult bighorn sheep ewes in the Ram Mountain population and the survival of lambs to one year of age, according to the number of years of data used for the correlation. With 15 years of data, coefficients of less than –0.51 are significant at $P = 0.05$.

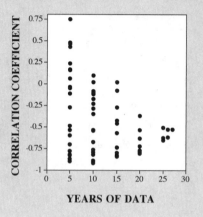

The scientific literature on reproductive strategies and population dynamics in other ungulates was of great benefit to our research. By comparing our results with those of other studies, we could better understand the ecology of mountain goats. Almost all the useful comparisons, however, were with a small number of other long-term studies of marked individuals that provide a disproportionate amount of available scientific information (table 13.1). The ability to monitor known-age individuals is a major advantage of long-term studies of marked ungulates, because in these species age has a strong effect on reproduction and survival (Côté and Festa-Bianchet 2001d; Coulson et al. 2001; Festa-Bianchet et al. 2003; Gaillard et al. 2001, 2003). Studies that mark most individuals in their research population also benefit from accurate yearly estimates of population parameters, including changes in sex–age structure. At Caw Ridge, an additional benefit of marking almost all individuals was the ability to identify the rare immigrants to the study area.

Long-term studies of ungulates also have limitations. They are costly, require accessible study areas and extensive field seasons every year, and the number of animals that can be marked and monitored by a research

TABLE 13.1
Long-term Studies of Marked Ungulates

Species	Study area	Country	Years	Predators?	Status	Major references
Alpine chamois	Les Bauges	France	1985–	No	Native	Loison et al. 1996
Alpine ibex	Belledonne	France	1985–	No	Reintroduced	Toïgo et al. 1999
Bighorn sheep	Sheep River	Canada	1981–	Yes	Native	Loison et al. 1999a
Bighorn sheep	Ram Mountain	Canada	1971–	Yes	Native	Coltman et al. 2003
Bighorn sheep	Bison Range[a]	USA	1979–	Yes	Introduced	Hogg and Forbes 1997
European mouflon	Caroux	France	1985–	No	Introduced	Cransac et al. 1997
Fallow deer	Phoenix Park[a]	Ireland	1985–	No[b]	Introduced	McElligott et al. 2002
Fallow deer	San Rossore	Italy	1984–	No	Introduced	Apollonio et al. 1998
Greater kudu	Kruger Park	South Africa	1984–1993	Yes	Native	Owen-Smith 1990
Moose	Vega	Norway	1992–	No	Native	Sæther et al. 2003
Mountain goat	Caw Ridge	Canada	1989–	Yes	Native	This study
Pronghorn antelope	Bison Range	USA	1981–	Yes	Introduced	Byers 1997
Pyrenean chamois	Orlu	France	1985–	No	Native	Loison et al. 1999a
Pyrenean chamois	Parc National des Pyrénées	France	1993–	No	Native	Crampe et al. 2002
Red deer	Rum	UK	1972–	No	Reintroduced	Clutton-Brock et al. 1982
Red deer	Trøndelag	Norway	1987–	No	Native	Langvatn and Loison 1999
Roe deer	Chizé[a]	France	1978	No	Native	Gaillard et al. 1997
Roe deer	Trois-Fontaines[a]	France	1976–	No	Native	Gaillard et al. 1997
Soay sheep	St. Kilda	UK	1985–	No	Introduced	Clutton-Brock et al. 1996

[a]Fenced population.
[b]Some mortality caused by domestic dogs, much mortality caused by vehicle collision.
Only studies lasting over 10 years and where more than 100 animals were marked are included. Predators are only those able to kill adult individuals.

team limits sample sizes. Very few have incorporated manipulations of density, food availability, predation, or reproductive effort (Jorgenson et al. 1993b; Tavecchia et al. 2005). Although they are often advocated, experimental manipulations tend to be incompatible with long-term monitoring of natural variability. A useful alternative is the analysis of data obtained from ungulates killed by sport hunters, which in some jurisdictions are carefully collected and organized (Milner et al. 2006; Pettorelli et al. 2007). The advantages of long-term data on harvested animals include a very large sample size, a wide geographical coverage (often including populations in very different environments), and the possibility to compare populations whose density and sex–age structures are experimentally manipulated through changes in hunting regulations (Mysterud et al. 2000, 2005a, b; Yoccoz et al. 2002). Studies that rely on harvested animals, however, have two major drawbacks. First, they cannot account for individual differences such as mass-specific survival or reproductive potential, because each individual is only sampled once. Second, and perhaps more importantly, they are subject to a potentially strong but often unquantified bias from hunter selectivity (Martinez et al. 2005; Mysterud et al. 2006), particularly where hunters are obliged by law to only harvest certain kinds of animals. For example, in most jurisdictions in North America it is illegal to harvest small-horned bighorn sheep rams, and rams with fast-growing horns are at risk of harvest at a younger age than rams with slow-growing horns (Coltman et al. 2003). In cross-sectional analyses of the population dynamics of hunted ungulates, the reluctance of hunters to kill young animals and females can strongly bias the estimation of vital rates. If these limitations and biases are ignored, analyses of data from harvested animals can easily lead to spurious conclusions. On the other hand, if the limitations are acknowledged and accounted for, studies based on harvested animals can provide strong contributions to our knowledge of wildlife ecology and our ability to manage wildlife species.

For most ungulate species, only one or two populations have been the subject of long-term monitoring (table 13.1). We do not know the extent to which one or two study populations may be representative of the species in general. Some results are affected by local conditions that may or may not be typical of most other populations, or even of the study population at different times. Mountain goats are no exception: currently, much of what is known about this species is based on information derived only from our study population.

Many of the study populations listed in table 13.1 are either fenced or isolated. Several studies monitored introduced or reintroduced populations, sometimes in habitats that are either radically altered or are not typical of the species. To overcome these limitations, researchers increasingly combine the results of several long-term studies, looking for common trends and trying to identify the sources of differences (Coulson et al. 2005; Festa-Bianchet et al. 2004; Gaillard et al. 2000b; Loison et al. 1999a). We expect that in the next few years there will be more research pooling data from several long-term studies, and comparing long-term studies of marked individuals with data from harvested animals.

The geographical distribution of long-term studies of ungulates is biased toward temperate and boreal environments (table 13.1). With the exception of greater kudu and reindeer, there are no published studies from arid, tropical, or arctic environments. In most study areas, large predators have been extirpated: the only intensively studied marked populations coexisting with a full natural complement of predators are the Caw Ridge mountain goats, kudu in South Africa, and the two bighorn populations in Alberta. Some studies are now documenting the consequences of reintroduction or natural recolonization by large carnivores (Berger et al. 2003). The population dynamics of ungulates can be very different in areas with and without predators (Owen-Smith et al. 2005; Sinclair et al. 2003). Because ungulates evolved in the presence of large predators, much of their behavior and possibly their reproductive strategies appear to be antipredator adaptations (Byers 1997; Festa-Bianchet 1991). Therefore one may question the generality of results obtained from predator-free areas, where some antipredator adaptations may lead to suboptimal behaviors (Byers 1997). Some demographic parameters, such as the survival of prime-aged adult females, appear to be very similar in ungulate populations with and without predators (Gaillard et al. 1998). For example, the survival of mountain goat females aged three to nine years on Caw Ridge, despite the presence of several species of large predators, is higher than the survival of prime-aged female ungulates in several predator-free populations. On the other hand, specialist predators can substantially lower the survival of adults of both sexes, as documented for bighorn sheep (Festa-Bianchet et al. 2006). These contrasting results underline the need for more research on ungulate populations exposed to large predators, particularly in view of the expanding range of some large predators in both North America and Europe.

There are essentially no published long-term data on individual survival and reproduction from marked populations of macropod marsupials. Information on the behavioral and population ecology of large macropods would provide opportunities to test some of the theories developed for placentals. Macropods may also provide new insights because of their many environmental, physiological, and reproductive differences compared to ungulates (Fisher et al. 2001; Johnson 1986a).

What Limits Long-Term Studies of Ungulates?

Long-term studies of ungulates require a strong personal commitment from investigators, stable funding, and some luck. The capture effort and regular monitoring of key biological events that characterize long-term studies have to be repeated every year. A one-year gap in the data will prevent the measurement of lifetime reproductive success, and the resulting incomplete individual life histories may be unsuitable for many analyses. Stable sources of funding are difficult to find, because few funding agencies appreciate the value of long-term research. Many funders experience study fatigue after a few years. Absurd comments such as "those animals have been studied to death" or "we already financed this last year" are often heard in reaction to requests for continued funding. Fortunately, in Canada the Natural Sciences and Engineering Research Council (NSERC) appreciates that long-term monitoring is required to understand ecology. NSERC support has been key to the Caw Ridge study, through operating grants and scholarships. Support from the Alberta Conservation Association is also contributing to the continuation of the project. Agencies that approach allocation of research funding based on short-term and shortsighted "problem-solving" are seldom interested in long-term research. That unproductive approach is surprising when one considers the importance of long-term monitoring for conservation and wildlife management. Fortunately, many biologists working for management agencies are highly supportive of long-term research and help long-term projects. The Caw Ridge study always enjoyed the support of Alberta Fish and Wildlife.

Finally, long-term research requires luck. At Caw Ridge, luck allowed us to continue accessing the study area by all-terrain vehicle even when much of the road was washed out by floods and landslides. Also, luck, mostly in the form of changes in the price of coal, has so far protected the study area from becoming an open-pit coal mine.

Ability to access the study area is essential, but limited access can also be an asset. Many study sites on table 13.1 are either on islands (Rum, St. Kilda, Svalbard) or mountains (Caw Ridge, Ram Mountain, Belledonne, Orlu, Les Bauges) or are fenced with controlled access (Bison Range, San Rossore, Chizé, Trois-Fontaines). A comparison of the three study areas in Alberta reveals the effects of excessive accessibility. Reaching the camp on Ram Mountain requires 35 km on a dirt road, 4 km by all-terrain vehicle, and a 1-km hike. The Sheep River study area, in contrast, is a one-hour drive on paved roads from Calgary, a city of 1 million people. The Sheep River bighorns have experienced habitat destruction through road construction, overgrazing by livestock, possible disease transmission from cattle, harassment by dogs, poaching, rock throwing by tourists, and occasional road kill. None of these problems has affected either Caw Ridge or Ram Mountain. At the other extreme, the fallow deer study at Phoenix Park has been highly successful despite being in a Dublin park, with cars and dogs as the main sources of mortality (McElligott et al. 2002, 2003).

The Value of Fundamental Research for Conservation and Wildlife Management

Long-term monitoring of unhunted populations provides an indication of what is normal, in terms of both estimating long-term means and quantifying temporal variability. That information constitutes a baseline to which managers can compare the effects of different harvest strategies, or assess the effects of human-induced habitat alterations (Sinclair 1991). The Caw Ridge study provided valuable information about yearly changes in sex–age structure and in population growth, which was incorporated in the provincial management strategy for mountain goats (Glasgow et al. 2003) and is now a reference point for mountain goat management over the species' entire range. For example, both the number and the proportion of adult males in the population changed substantially during the study. Any two- or three-year window would likely have not led to the same conclusions about mountain goat ecology and conservation that we can draw after fifteen years of research (box 13.1). Management actions based on short-term data can be unjustified. The study population was declining or stagnant between 1989 and 1992, yet it increased in almost every one of the following ten years. Had we started a predator control or habitat-manipulation program (such as prescribed fire, or chemical fertilization) in the early 1990s, we could have claimed a huge success!

Although long-term studies of marked ungulates (table 13.1) have provided much valuable information for conservation and wildlife management, very few were conducted on hunted populations. Among most of those that were hunted or subject to regular removals, marked animals typically enjoyed special protection, so that they were not subject to the same mortality as unmarked animals. For the two studies of bighorn sheep in Alberta, hunting was a major cause of death for adult males (Festa-Bianchet 1989b; Jorgenson et al. 1993b), and one population was manipulated to simulate a ewe hunting season (Jorgenson et al. 1993b). Only one study of red deer in Norway (Langvatn and Loison 1999) was conducted on a population with substantial hunting mortality for both sexes. Consequently, very little information is available on age-specific survival and other life-history traits of marked individuals in hunted populations. Studies based on harvest records strongly suggest that in hunted populations life expectancy is reduced substantially, especially for males (Mysterud et al. 2005b), and that the age- and sex-specific mortality patterns are very different from those seen in unhunted populations (Wright et al. 2006).

Future Research on Caw Ridge

The study of mountain goats on Caw Ridge is now expanding into population genetics, selection, and the effects of changes in density. One gaping hole in our research, which we are currently addressing, is the lack of information on male reproductive success. Over the last twelve years we have collected tissue samples from most marked goats, and we are now beginning to identify paternities through DNA analyses, as has been recently done for a few other ungulates (Coltman et al. 1999a, 2002; Pemberton et al. 1992). We expect both age and body mass to play major roles in male reproductive success, but we do not expect a strong effect of horn size (chapter 6). Because we do not know what may be the possible effects of female choice on male reproductive success (Balmford et al. 1992; Byers et al. 1994; Hogg 1987), we expect some surprises.

The synchrony of births (fig. 7.4) and our preliminary data during the rut suggest that most estruses occur over a short period. Therefore, dominant males may have a limited ability to monopolize matings, because the mating system apparently involves defense of a single estrous female at a time (Geist 1964). Given the scarcity of mature males, it will be important to determine if there is any evidence of incest avoidance (Berger and Cunningham 1987; Clutton-Brock 1989a; Coulson et al. 1998), and to compare the reproductive success of immigrant and phylopatric males. The pattern of male reproductive success will likely also provide some very useful information concerning the possible effects of sport harvest of adult males.

Future analyses of genotype, morphology, and life history of both sexes will investigate the possible interactions of heterozygosity, resource availability, and previous life history on the age-specific survival and reproductive success of individuals. Because we have sampled about 98% of individuals for DNA in the last ten years, we will be able to test the effects of inbreeding on reproductive performance by quantifying inbreeding based on pedigree data, which are very difficult to obtain for most wild mammals. Inbreeding rates estimated by sampling methods such as mean heterozygosity are easier to obtain because not all animals need to be sampled, but their reliability is lower than that of coefficients calculated from pedigrees (Keller and Waller 2002). Very few studies have sampled enough individuals to determine complete pedigrees with a high degree of certainty (Marshall et al. 2002). Our data base will also allow us to calculate selection coefficients of various life-history traits.

The use of remote-controlled platform scales (Bassano et al. 2003) has opened a number of new research avenues that require repeated measures of individual body mass. In particular, we will be able to measure any somatic costs of reproduction (Festa-Bianchet et al. 1995, 1998), examine the possible effects of genetic variability on mass and mass gain patterns, and further explore the effects of mass on life-history traits.

Finally, our long-term database on known individuals offers an exciting opportunity to assess the possible effects of global warming. Alpine environments are expected to change substantially as the climate warms, and already we are seeing indications that the parturition season may be earlier than fifteen years ago. Continued monitoring of reproductive events and of body and horn growth will allow us to assess how mountain goats may be affected by climate change.

The Future of Caw Ridge and Its Mountain Goats

Despite a substantial decline in numbers in the southern parts of their historical range, mountain goats are not an endangered species. Where they are hunted, harvest rates are generally sustainable (table 12.1), and managers respond to population declines by lowering quotas or closing hunting seasons. Like most animals dependent on alpine ecosystems, mountain goats face an uncertain future because of global warming (Hughes 2000; Laurance 2001; Thomas et al. 2004). In addition, the Caw Ridge population faces two threats: coal mining and increased motorized access. Of these, mining is the greatest.

Caw Ridge has substantial coal deposits. When we began our research, the chances of a long-term study seemed remote, because of a plan to turn the eastern section of the study area (including the location of our cabin and the traps) into an open-pit coal mine. That plan was shelved, mostly because of economic concerns and changes in foreign demand for coal, but not completely abandoned. Mining operations ceased in March 2000 but reopened in spring of 2004 as the price of coal increased. Current plans, pending approval by the Alberta government, are to mine within 3 km of our trap site and then expand close to the east side of the west end (fig. 2.6), eventually obliterating about 80% of the study area. At the same time, forestry operations to the west and north of

Caw Ridge are expanding. More clearcuts are visible from the ridge each year. Although unlikely to have a direct impact on mountain goats, clearcuts have a negative effect on the caribou that migrate through Caw Ridge (Smith et al. 2000). The effects of forestry on wildlife include altered predator–prey relationships (James and Stuart-Smith 2000; Kinley and Apps 2001; Schaefer 2003; Terry et al. 2000), and their consequences for adjacent populations of mountain goats (that may become alternative prey for increased predator populations) are unpredictable. Mostly because of expanding forestry activities, woodland caribou are a threatened species in Canada (COSEWIC 2002). That federal status, however, has so far been ineffective in protecting their habitat in areas controlled by provincial governments.

Access to the study area is essential for our long-term research. While currently not a major threat to mountain goats, increased motorized access to Caw Ridge may affect both the goats and their habitat in the future. In most weekends in July and August, over twenty ATVs now ride over Caw Ridge and the number recorded during our summer field season has increased from a few tens to several hundreds in less than ten years. Their impact is worsened by the lack of regulations on off-road access. Although most remain on the main trails, some ride off-trail, damaging the fragile alpine tundra and leaving erosion sites that take years to recover. Erosion left by off-trail use ten years ago is still clearly visible today. Off-trail riding also makes ATVs more unpredictable for mountain goats, and we witnessed a few cases of willful harassment of goats. Conservation of wildlife habitat on Caw Ridge will require that motorized access be restricted to specific trails. Signs encouraging ATV users to remain on the trail have recently been posted, but effective enforcement will also be required.

Caw Ridge is a unique alpine ecosystem. It provides habitat to eleven species of large mammals, including a large herd of mountain goats, one of the last two migratory herds of woodland caribou in Alberta, and grizzly bears, considered threatened in the province. It deserves protection for our benefit and that of future generations. Its inclusion within the Willmore Wilderness Area would be a positive step, but alternative ways to protect it are also available, such as declaring Caw Ridge a wildlife sanctuary or simply prohibiting motorized access and habitat destruction by using the provisions of existing laws. What is required is the political will to act, and we will work hard to stimulate that action.

Appendix

Mammals and birds seen on Caw Ridge during the study.

Mammals

Artodactila
Moose, *Alces alces*
White-tailed deer, *Odocoileus virginianus*
Mule deer, *Odocoileus hemionus*
Wapiti, *Cervus elaphus*
Caribou, *Rangifer tarandus*
Mountain goat, *Oreamnos americanus*
Bighorn sheep, *Ovis canadensis*

Rodentia

Hoary marmot, *Marmota caligata*
Columbian ground squirrel, *Spermophilus columbianus*
Golden-mantled ground squirrel, *Spermophilus lateralis*
Red squirrel, *Tamiasciurus hudsonicus*
Least chipmunk, *Eutamias minimus*
Bushy-tailed woodrat, *Neotoma cinerea*
Deermouse, *Peromyscus maniculatus*
Porcupine, *Erethizon dorsatum*
Beaver, *Castor canadensis*

Lagomorpha

Pika, *Ochotona princeps*
Snowshoe hare, *Lepus americanus*

Carnivora

Grizzly bear, *Ursus arctos*
Black bear, *Ursus americanus*
Wolf, *Canis lupus*
Coyote, *Canis latrans*
Red fox, *Vulpes vulpes*
Cougar, *Puma concolor*
Lynx, *Lynx canadensis*
Wolverine, *Gulo gulo*
Longtail weasel, *Mustela frenata*

Birds

Canada goose, *Branta canadensis*
Mallard, *Anas platyrhynchos*
Cooper's hawk, *Accipiter cooperi*
Northern harrier, *Circus cyaneus*
Red-tailed hawk, *Buteo jamaicensis*
Rough-legged hawk, *Buteo lagopus*
Golden eagle, *Aquila chrysaetos*
Bald eagle, *Haliaeetus leucocephalus*
Osprey, *Pandion haliaetus*
Prairie falcon, *Falco mexicanus*
Peregrine falcon, *Falco peregrinus*
American kestrel, *Falco sparverius*
Blue grouse, *Dendragapus obscurus*
Ruffed grouse, *Bonasa umbellus*
Spruce grouse, *Falcipennis canadensis*
White-tailed ptarmigan, *Lagopus leucurus*
American golden-plover, *Pluvialis dominica*
Killdeer, *Charadrius vociferus*
Short-eared owl, *Asio flammeus*
Rufus hummingbird, *Selapshorus rufus*
Northern flicker, *Colaptes auratus*
Horned lark, *Eremophila alpestris*

Barn swallow, *Hirundo rustica*
Grey jay, *Perisoreus canadensis*
Clark's nutracker, *Nucifraga columbiana*
Common raven, *Corvus corax*
Black-capped chickadee, *Parus atricapillus*
Mountain chickadee, *Parus gambeli*
Boreal chickadee, *Parus hudsonicus*
Townsend's solitaire, *Myadestes townsendi*
Mountain bluebird, *Sialia currucoides*
Audubon's warbler, *Dendroica auduboni*
Pine grosbeak, *Pinicola enucleator*
Grey-crowned rosy finch, *Leucosticte tephrocotis*
White-crowned sparrow, *Zonotrichia leucophrys*
Golden-crowned sparrow, *Zonotrichia atricapilla*
Say's phoebe, *Sayornis saya*
American robin, *Turdus migratorius*
American pipit, *Anthrus rubescens*
Cedar waxwing, *Bombycilla cedrorum*
Sandpiper spp., *Calidris spp.*

Scientific Names of Other Species Mentioned in the Text

Alpine chamois, *Rupicapra rupicapra*
Alpine ibex, *Capra ibex*
Bison, *Bison bison*
Cuvier's gazelle, *Gazella cuvieri*
European mouflon, *Ovis aries*
Fallow deer, *Dama dama*
Goral, *Naemorhedus* spp.
Greater kudu, *Tragelaphus strepsiceros*
Isard or Pyrenean chamois, *Rupicapra pirenaica*
Japanese serow, *Capricornis crispus*
Mainland serow, *Capricornis sumatraensis*
Muskox, *Ovibos moschatus*
Nubian ibex, *Capra nubiana*
Pronghorn antelope, *Antilocapra americana*
Red deer, *Cervus elaphus*
Roe deer, *Capreolus capreolus*
Sable antelope, *Hyppotragus niger*
Thinhorn (Dall's) sheep, *Ovis dalli*
Wild boar, *Sus scrofa*

Literature Cited

Adams, K. P., and P. J. Pekins. 1995. Growth patterns of New England moose: yearlings as indicators of population status. *Alces* 31:53–59.

Adams, L. G., and J. A. Bailey. 1982. Population dynamics of mountain goats in the Sawatch Range, Colorado. *Journal of Wildlife Management* 46:1003–1009.

———. 1983. Winter forages of mountain goats in central Colorado. *Journal of Wildlife Management* 47:1237–1243.

Adams, L. G., and B. W. Dale. 1998. Reproductive performance of female Alaskan caribou. *Journal of Wildlife Management* 62:1184–1195.

Adams, L. G., M. A. Masteller, and J. A. Bailey. 1982. Movements and home range of mountain goats, Sheep Mountain–Gladstone Ridge, Colorado. *Proceedings of the Biennial Symposium of the Northern Wild Sheep & Goat Council* 3:391–405.

Alados, C. L., and J. M. Escos. 1994. Variation in the sex ratio of a low dimorphic polygynous species with high levels of maternal reproductive effort: Cuvier's gazelle. *Ethology Ecology and Evolution* 6:301–311.

Albon, S. D., H. J. Staines, F. E. Guinness, and T. H. Clutton-Brock. 1992. Density-dependent changes in the spacing behaviour of female kin in red deer. *Journal of Animal Ecology* 61:131–137.

Altmann, J. 1974. Observational study of behavior: sampling methods. *Behaviour* 49:227–267.

Alvarez, F. 1990. Horns and fighting in male Spanish ibex, *Capra pyrenaica*. *Journal of Mammalogy* 71:608–616.

———. 1994. Bone density and breaking stress in relation to consistent fracture position in fallow deer antlers. *Doñana, Acta Vertebrata* 21:15–24.

Apollonio, M., B. Bassano, and A. Mustoni. 2003. Behavioral aspects of conservation and management of European mammals. 157–170 in *Animal behavior and wildlife conservation*, ed. M. Festa-Bianchet, M. Apollonio. Washington, DC: Island Press.

Apollonio, M., S. Focardi, S. Toso, and L. Nacci. 1998. Habitat selection and group formation patterns of fallow deer *Dama dama* in a submediterranean environment. *Ecography* 21:225–234.

Arnold, S. J., and M. J. Wade. 1984. On the measurement of natural and sexual selection: applications. *Evolution* 38:720–734.

Bailey, J. A. 1991. Reproductive success in female mountain goats. *Canadian Journal of Zoology* 69:2956–2961.

Ballard, W. B., H. A. Whitlaw, B. F. Wakeling, R. L. Brown, J. C. deVos, and M. C. Wallace. 2000. Survival of female elk in northern Arizona. *Journal of Wildlife Management* 64:500–504.

Balmford, A. P., A. M. Rosser, and S. D. Albon. 1992. Correlates of female choice in resource-defending antelope. *Behavioral Ecology and Sociobiology* 31:107–114.

Bartmann, R. M., G. C. White, and L. H. Carpenter. 1992. Compensatory mortality in a Colorado mule deer population. *Wildife Monographs* 121:1–39.

Bassano, B., A. v. Hardenberg, F. Pelletier, and G. Gobbi. 2003. A method to weigh free-ranging ungulates without handling. *Wildlife Society Bulletin* 31:1205–1209.

Beckerman, A., T. G. Benton, E. Ranta, V. Kaitala, and P. Lundberg. 2002. Population dynamic consequences of delayed life-history effects. *Trends in Ecology and Evolution* 17:263–269.

Beissinger, S. R., and M. I. Westphal. 1998. On the use of demographic models of population viability in endangered species management. *Journal of Wildlife Management* 62:821–841.

Benton, T. G., and A. Grant. 1999. Elasticity analysis as an important tool in evolutionary and population ecology. *Trends in Ecology and Evolution* 14:467–471.

Benton, T. G., A. Grant, and T. H. Clutton-Brock. 1995. Does environmental stochasticity matter? Analysis of red deer life-histories on Rum. *Evolutionary Ecology* 9:559–574.

Berger, J. 1992. Facilitation of reproductive synchrony by gestation adjustment in gregarious mammals: a new hypothesis. *Ecology* 73:323–329.

Berger, J., and C. Cunningham. 1987. Influence of familiarity on frequency of inbreeding in wild horses. *Evolution* 41:229–231.

Berger, J., S. L. Monfort, T. Roffe, P. B. Stacey, and J. W. Testa. 2003. Through the eyes of prey: how the extinction and conservation of North America's large carnivores alter prey systems and biodiversity. 133–156 in *Animal behavior and wildlife conservation*, ed. M. Festa-Bianchet, M. Apollonio. Washington, DC: Island Press.

Berger, J., J. E. Swenson, and I. L. Persson. 2001. Recolonizing carnivores and naive prey: conservation lessons from Pleistocene extinctions. *Science* 291:1036–1039.

Bergerud, A. T., and W. B. Ballard. 1988. Wolf predation on caribou: the Nelchina case history, a different interpretation. *Journal of Wildlife Management* 52:344–357.

Bergerud, A. T., R. Ferguson, and H. E. Butler. 1990. Spring migration and dispersal of woodland caribou at calving. *Animal Behaviour* 39:360–368.

Bérubé, C., M. Festa-Bianchet, and J. T. Jorgenson. 1999. Individual differences, longevity, and reproductive senescence in bighorn ewes. *Ecology* 80:2555–2565.

Bérubé, C. H. 1997. *Les stratégies d'adaptation vitale chez les brebis du mouflon d'Amérique* (Ovis canadensis): *la reproduction en fonction de l'âge*. Ph.D. thesis, Université de Sherbrooke.

Bérubé, C. H., M. Festa-Bianchet, and J. T. Jorgenson. 1996. Reproductive costs of sons and daughters in Rocky Mountain bighorn sheep. *Behavioral Ecology* 7:60–68.

Birgersson, B. 1998a. Adaptive adjustment of the sex ratio: more data and considerations from a fallow deer population. *Behavioral Ecology* 9:404–408.

———. 1998b. Male-biased maternal expenditure and associated costs in fallow deer. *Behavioral Ecology and Sociobiology* 43:87–93.

Birgersson, B., and K. Ekvall. 1997. Early growth in male and female fallow deer fawns. *Behavioral Ecology* 8:493–499.

Blanchard, P. 2002. *Survivre pour se reproduire: rôle de la condition individuelle dans les tactiques de reproduction chez le mouflon américain* (Ovis canadensis). Ph.D. thesis, Université de Sherbrooke.

Blanchard, P., M. Festa-Bianchet, J.-M. Gaillard, and J. T. Jorgenson. 2003. A test of long-term fecal nitrogen monitoring to evaluate nutritional status in bighorn sheep. *Journal of Wildlife Management* 67:477–484.

———. 2004. Maternal condition and offspring sex ratio in polygynous ungulates: a case study of bighorn sheep. *Behavioral Ecology* 16:274–279.

Bleich, V. C., R. T. Bowyer, and J. D. Wehausen. 1997. Sexual segregation in mountain sheep: resources or predation? *Wildlife Monographs* 134:1–50.

Bon, R., C. Rideau, J. C. Villaret, and J. Joachim. 2001. Segregation is not only a matter of sex in Alpine ibex, *Capra ibex ibex*. *Animal Behaviour* 62:495–504.

Bonenfant, C., L. E. Loe, A. Mysterud, R. Langvatn, N. C. Stenseth, J. M. Gaillard, and F. Klein. 2004. Multiple causes of sexual segregation in European red deer: enlightenments from varying breeding phenology at high and low latitude. *Proceedings of the Royal Society of London B* 271:883–892.

Boschi, C., and B. Nievergelt. 2003. The spatial patterns of Alpine chamois (*Rupicapra rupicapra rupicapra*) and their influence on population dynamics in the Swiss National Park. *Mammalian Biology* 64:16–30.

Boulinier, T., G. Sorci, J. Clobert, and E. Danchin. 1997. An experimental study of the costs of reproduction in the kittiwake *Rissa tridactyla*: comment. *Ecology* 78:1284–1287.

Boutin, S. 1992. Predation and moose population dynamics: a critique. *Journal of Wildlife Management* 56:116–127.

Bowyer, R. T. 2005. Sexual segregation in ruminants: definitions, hypotheses, and implications for conservation and management. *Journal of Mammalogy* 85:1039–1052.

Bowyer, R. T., V. v. Ballenberghe, and J. G. Kie. 1998. Timing and synchrony of parturition in Alaskan moose: long-term versus proximal effects of climate. *Journal of Mammalogy* 79:1332–1344.

Boyce, M. S. 1989. *The Jackson elk herd*. Cambridge: Cambridge University Press.

Boyce, M. S., A. R. E. Sinclair, and G. C. White. 1999. Seasonal compensation of predation and harvesting. *Oikos* 87:419–426.

Brandborg, S. M. 1955. Life history and management of the mountain goat in Idaho. *Idaho Wildlife Bulletin* 2:1–142.

Brodie, E. D., and F. J. Janzen. 1996. On the assignment of fitness values in statistical analyses of selection. *Evolution* 50:437–442.

Bryant, J. P., F. D. Provenza, J. Pastor, P. B. Reichardt, T. P. Clausen, and J. T. du Toit. 1991. Interactions between woody plants and browsing mammals mediated by secondary metabolites. *Annual Review of Ecology and Systematics* 22:431–446.

Bunnell, F. L. 1978. Horn growth and population quality in Dall sheep. *Journal of Wildlife Management* 42:764–775.

———. 1980. Factors controlling lambing period of Dall's sheep. *Canadian Journal of Zoology* 58:1027–1031.

———. 1982. The lambing period of mountain sheep: synthesis, hypotheses, and tests. *Canadian Journal of Zoology* 60:1–14.

Burles, D. W., and M. Hoefs. 1984. Winter mortality of Dall sheep, *Ovis dalli dalli*, in Kluane National Park, Yukon. *Canadian Field-Naturalist* 98:479–484.

Byers, J. A. 1997. *American pronghorn*. Chicago: University of Chicago Press.

Byers, J. A., and J. T. Hogg. 1995. Environmental effects on prenatal growth rate in pronghorn and bighorn: further evidence for energy constraint on sex-biased maternal expenditure. *Behavioral Ecology* 6:451–457.

Byers, J. A., and D. W. Kitchen. 1988. Mating system shift in a pronghorn population. *Behavioral Ecology and Sociobiology* 22:355–360.

Byers, J. A., J. D. Moodie, and N. Hall. 1994. Pronghorn females choose vigorous mates. *Animal Behaviour* 47:33–43.

Cabana, G., and D. L. Kramer. 1991. Random offspring mortality and variation in parental fitness. *Evolution* 45:228–234.

Cam, E., W. A. Link, E. G. Cooch, J. Y. Monnat, and E. Danchin. 2002. Individual covariation in life-history traits: seeing the trees despite the forest. *American Naturalist* 159:96–105.

Cameron, E. Z., W. L. Linklater, K. J. Stafford, and E. O. Minot. 2000. Aging and improving reproductive success in horses: declining residual reproductive value or just older and wiser? *Behavioral Ecology and Sociobiology* 47:243–249.

Cameron, R. D., W. T. Smith, S. G. Fancy, K. L. Gerhart, and R. G. White. 1993. Calving success of female caribou in relation to body weight. *Canadian Journal of Zoology* 71:480–486.

Carl, G. R., and C. T. Robbins. 1988. The energetic cost of predator avoidance in neonatal ungulates: hiding versus following. *Canadian Journal of Zoology* 66:239–246.

Caswell, H. 2001. *Matrix population models: construction, analysis, and interpretation.* Sunderland, MA: Sinauer.

Catchpole, E. A., Y. Fan, B. J. T. Morgan, T. H. Clutton-Brock, and T. Coulson. 2004. Sexual dimorphism, survival and dispersal in red deer. *Journal of Agriculture, Biology and Environmental Statistics* 9:1–26.

Caughley, G. 1970. Eruption of ungulate populations, with emphasis on Himalayan thar in New Zealand. *Ecology* 51:53–72.

———. 1994. Directions in conservation biology. *Journal of Animal Ecology* 63:215–244.

Caughley, G., and A. R. E. Sinclair. 1994. *Wildlife ecology and management.* Boston: Blackwell Scientific Publications.

Chadwick, D. H. 1977. The influence of mountain goat social relationships on population size and distribution. *Proceedings of the First International Mountain Goat Symposium* 1:74–91.

Chan-McLeod, A. C., R. G. White, and D. E. Russell. 1999. Comparative body composition strategies of breeding and nonbreeding female caribou. *Canadian Journal of Zoology* 77:1901–1907.

Clover, M. R. 1956. Single-gate deer trap. *California Fish and Game* 42:199–201.

Clutton-Brock, T. H. 1982. The functions of antlers. *Behaviour* 79:109–125.

———. 1989a. Female transfer and inbreeding avoidance in social mammals. *Nature* 337:70–72.

———. 1989b. Mammalian mating systems. *Proceedings of the Royal Society of London B* 236:339–372.

———. 1991. *The evolution of parental care.* Princeton: Princeton University Press.

Clutton-Brock, T. H., S. D. Albon, and F. E. Guinness. 1981. Parental investment in male and female offspring in polygynous mammals. *Nature* 289:487–489.

———. 1984. Maternal dominance, breeding success and birth sex ratios in red deer. *Nature* 308:358–360.

———. 1985. Parental investment and sex differences in juvenile mortality in birds and mammals. *Nature* 313:131–133.

———. 1986. Great expectations: dominance, breeding success and offspring sex ratio in red deer. *Animal Behaviour* 34:460–471.

Clutton-Brock, T. H., and T. Coulson. 2002. Comparative ungulate dynamics: the devil is in the detail. *Philosophical Transactions of the Royal Society B* 357:1285–1298.

Clutton-Brock, T. H., D. Green, M. Hiraiwa-Hasegawa, and S. D. Albon. 1988. Passing the buck: resource defence, lek breeding and mate choice in fallow deer. *Behavioral Ecology and Sociobiology* 23:281–296.

Clutton-Brock, T. H., F. E. Guinness, and S. D. Albon. 1982. *Red deer: behavior and ecology of two sexes.* Chicago: University of Chicago.

———. 1983. The costs of reproduction to red deer hinds. *Journal of Animal Ecology* 52:367–383.

Clutton-Brock, T. H., and G. R. Iason. 1986. Sex ratio variation in mammals. *Quarterly Review of Biology* 61:339–374.

Clutton-Brock, T. H., G. R. Iason, and F. E. Guinness. 1987. Sexual segregation and density-related changes in habitat use in male and female red deer (*Cervus elaphus*). *Journal of Zoology* 211:275–289.

Clutton-Brock, T. H., A. W. Illius, K. Wilson, B. T. Grenfell, A. D. C. MacColl, and S. D. Albon. 1997a. Stability and instability in ungulate populations: an empirical analysis. *American Naturalist* 149:195–219.

Clutton-Brock, T. H., and K. McComb. 1993. Experimental tests of copying and mate choice in fallow deer (*Dama dama*). *Behavioral Ecology* 4:191–193.

Clutton-Brock, T. H., K. E. Rose, and F. E. Guinness. 1997b. Density-related changes in sexual selection in red deer. *Proceedings of the Royal Society of London B* 264:1509–1516.

Clutton-Brock, T. H., I. R. Stevenson, P. Marrow, A. D. MacColl, A. I. Houston, and J. M. McNamara. 1996. Population fluctuations, reproductive costs and life-history tactics in female Soay sheep. *Journal of Animal Ecology* 65:675–689.

Coggins, V. L. 2002. Status of Oregon Rocky Mountain Goats. *Proceedings of the Northern Wild Sheep and Goat Council* 13:62–67.

Coltman, D. W., D. R. Bancroft, A. Robertson, J. A. Smith, T. H. Clutton-Brock, and J. M. Pemberton. 1999a. Male reproductive success in a promiscuous mammal: behavioural estimates compared with genetic paternity. *Molecular Ecology* 8:1199–1209.

Coltman, D. W., M. Festa-Bianchet, J. T. Jorgenson, and C. Strobeck. 2002. Age-dependent sexual selection in bighorn rams. *Proceedings of the Royal Society of London B* 269:165–172.

Coltman, D. W., P. O'Donoghue, J. T. Jorgenson, J. T. Hogg, and M. Festa-Bianchet. 2005. Selection and genetic (co)variance in bighorn sheep. *Evolution* 59:1372–1382.

Coltman, D. W., P. O'Donoghue, J. T. Jorgenson, J. T. Hogg, C. Strobeck, and M. Festa-Bianchet. 2003. Undesirable evolutionary consequences of trophy hunting. *Nature* 426:655–658.

Coltman, D. W., J. A. Smith, D. R. Bancroft, J. Pilkington, A. D. C. MacColl, T. H. Clutton-Brock, and J. M. Pemberton. 1999b. Density-dependent variation in lifetime breeding success and natural and sexual selection in Soay rams. *American Naturalist* 154:730–746.

Conradt, L. 1998. Could asynchrony in activity between the sexes cause intersexual social segregation in ruminants? *Proceedings of the Royal Society of London B* 265:1–5.

Conradt, L., T. H. Clutton-Brock, and D. Thomson. 1999. Habitat segregation in ungulates: are males forced into suboptimal foraging habitats through indirect competition by females? *Oecologia* 119:367–377.

COSEWIC. 2002. COSEWIC assessment and update Status Report on the woodland caribou *Rangifer tarandus caribou* in Canada. Ottawa: Committee on the Status of Endangered Wildlife in Canada; xi + 98 pp.

Côté, S. D. 1996. Mountain goat responses to helicopter disturbance. *Wildlife Society Bulletin* 24:681–685.

———. 1999. *Dominance sociale et traits d'histoire de vie chez les femelles de la chèvre de montagne.* Ph.D. thesis, Université de Sherbrooke.

———. 2000. Dominance hierarchies in female mountain goats: stability, aggressiveness and determinants of rank. *Behaviour* 137:1541–1566.

Côté, S. D., and C. Beaudoin. 1997. Grizzly bear (*Ursus arctos*) attacks and nanny–kid separation on mountain goats (*Oreamnos americanus*). *Mammalia* 61:614–617.

Côté, S. D., and M. Festa-Bianchet. 2001a. Birthdate, mass and survival in mountain goat kids: effects of maternal characteristics and forage quality. *Oecologia* 127:230–238.

———. 2001b. Life-history correlates of horn asymmetry in mountain goats. *Journal of Mammalogy* 82:389–400.

———. 2001c. Offspring sex ratio in relation to maternal age and social rank in mountain goats. *Behavioral Ecology and Sociobiology* 49:260–265.

———. 2001d. Reproductive success in female mountain goats: the influence of maternal age and social rank. *Animal Behaviour* 62:173–181.

———. 2003. Mountain goat, *Oreamnos americanus.* 1061–1075 in *Wild mammals of North America: biology, management, conservation*, ed. G. A. Feldhamer, B. Thompson, J. Chapman. Baltimore: John Hopkins University Press.

Côté, S. D., M. Festa-Bianchet, and F. Fournier. 1998a. Life-history effects of chemical immobilization and radio collars in mountain goats. *Journal of Wildlife Management* 62:745–
752.

Côté, S. D., M. Festa-Bianchet, and K. G. Smith. 1998b. Horn growth in mountain goats
(*Oreamnos americanus*). *Journal of Mammalogy* 79:406–414.

———. 2001. Compensatory reproduction in harvested mountain goat populations: a word
of caution. *Wildlife Society Bulletin* 29:726–730.

Côté, S. D., A. Peracino, and G. Simard. 1997a. Wolf (*Canis lupus*) predation and maternal de-
fensive behavior in mountain goat (*Oreamnos americanus*). *Canadian Field-Naturalist*
111:389–392.

Côté, S. D., T. P. Rooney, J. P. Tremblay, C. Dussault, and D. M. Waller. 2004. Ecological impacts of deer overabundance. *Annual Review of Ecology, Evolution and Systematics* 35:113–
147.

Côté, S. D., J. A. Schaefer, and F. Messier. 1997b. Time budgets and synchrony of activities in
muskoxen: the influence of sex, age, and season. *Canadian Journal of Zoology* 75:1628–
1635.

Coulson, T., T. G. Benton, P. Lundberg, S. R. X. Dall, B. E. Kendall, and J. M. Gaillard.
2006. Estimating individual contributions to population growth: evolutionary fitness in
ecological time. *Proceedings of the Royal Society of London B* 273:547–555.

Coulson, T., E. A. Catchpole, S. D. Albon, B. J. T. Morgan, J. M. Pemberton, T. H. Clutton-
Brock, M. J. Crawley, and B. T. Grenfell. 2001. Age, sex, density, winter weather, and
population crashes in Soay sheep. *Science* 292:1528–1531.

Coulson, T., J. M. Gaillard, and M. Festa-Bianchet. 2005. Decomposing the variation in population growth into contributions from multiple demographic rates. *Journal of Animal
Ecology* 74:789–801.

Coulson, T., F. Guinness, J. Pemberton, and T. Clutton-Brock. 2004. The demographic consequences of releasing a population of red deer from culling. *Ecology* 85:411–422.

Coulson, T., L. E. B. Kruuk, G. Tavecchia, J. M. Pemberton, and T. H. Clutton-Brock. 2003.
Estimating selection on neonatal traits in red deer using elasticity path analysis. *Evolution*
57:2879–2892.

Coulson, T., E. J. Milner-Gulland, and T. H. Clutton-Brock. 2000. The relative roles of density and climatic variation on population dynamics and fecundity rates in three contrasting ungulate species. *Proceedings of the Royal Society of London B* 1454:1771–1779.

Coulson, T. N., J. M. Pemberton, S. D. Albon, M. Beaumont, T. C. Marshall, J. Slate, F. E.
Guinness, and T. H. Clutton-Brock. 1998. Microsatellites reveal heterosis in red deer.
Proceedings of the Royal Society of London B 265:489–495.

Cowan, I. M., and W. McCrory. 1970. Variation in the mountain goat, *Oreamnos americanus*
(Blainville). *Journal of Mammalogy* 51:60–73.

Crampe, J. P., J.-M. Gaillard, and A. Loison. 2002. L'enneigement hivernal: un facteur de
variation de recruetment chez l'isard (*Rupicapra pyrenaica pyrenaica*). *Canadian Journal of
Zoology* 80:1306–1312.

Cransac, N., A. J. M. Hewison, J. M. Gaillard, J. M. Cugnasse, and M. L. Maublanc. 1997.
Patterns of mouflon (*Ovis gmelini*) survival under moderate environmental conditions: effects of sex, age, and epizootics. *Canadian Journal of Zoology* 75:1867–1875.

Crête, M., and R. Courtois. 1997. Limiting factors might obscure population regulation of
moose (Cervidae: *Alces alces*) in unproductive boreal forest. *Journal of Zoology* 242:765–
781.

Crête, M., J. Huot, R. Nault, and R. Patenaude. 1993. Reproduction, growth and body composition of Rivière George caribou in captivity. *Arctic* 46:189–196.

Dailey, T. V., N. T. Hobbs, and T. N. Woodard. 1984. Experimental comparisons of diet selection by mountain goats and mountain sheep in Colorado. *Journal of Wildlife Management* 48:799–806.

Dauphiné, T. C., and R. L. McClure. 1974. Synchronous mating in barren-ground caribou. *Journal of Wildlife Management* 38:54–66.

DelGiudice, G. D., R. O. Peterson, and W. M. Samuel. 1997. Trends of winter nutritional restriction, ticks, and numbers of moose on Isle Royale. *Journal of Wildlife Management* 61:895–903.

Dobson, F. S. 1982. Competition for mates and predominant juvenile male dispersal in mammals. *Animal Behaviour* 30:1183–1192.

Dobson, F. S., T. S. Risch, and J. O. Murie. 1999. Increasing returns in the life history of Columbian ground squirrels. *Journal of Animal Ecology* 68:73–86.

Dumont, A., M. Crête, J. P. Ouellet, J. Huot, and J. Lamoureux. 2000. Population dynamics of northern white-tailed deer during mild winters: evidence of regulation by food competition. *Canadian Journal of Zoology* 78:764–776.

Edmonds, E. J. 1988. Population status, distribution, and movements of woodland caribou in west central Alberta. *Canadian Journal of Zoology* 66:817–826.

Enck, J. W., D. J. Decker, and T. L. Brown. 2000. Status of hunter recruitment and retention in the United States. *Wildlife Society Bulletin* 28:817–824.

Enggist-Düblin, P., and P. Ingold. 2003. Modelling the impact of different forms of wildlife harassment, exemplified by a quantitative comparison of the effects of hikers and paragliders on feeding and space use of chamois *Rupicapra rupicapra*. *Wildlife Biology* 9:37–45.

Ericsson, G., K. Wallin, J. P. Ball, and M. Broberg. 2001. Age-related reproductive effort and senescence in free-ranging moose, *Alces alces*. *Ecology* 82:1613–1620.

Ernest, H. B., E. S. Rubin, and W. M. Boyce. 2002. Fecal DNA analysis and risk assessment of mountain lion predation of bighorn sheep. *Journal of Wildlife Management* 66:75–85.

Estes, R. D. 1976. The significance of breeding synchrony in the wildebeest. *East African Wildlife Journal* 14:135–152.

———. 1991. The significance of horns and other male secondary sexual characters in female bovids. *Applied Animal Behaviour Science* 29:403–451.

Fairbanks, W. S. 1993. Birthdate, birthweight, and survival in pronghorn fawns. *Journal of Mammalogy* 74:129–135.

Fandos, P. 1995. Factors affecting horn growth in male Spanish ibex (*Capra pyrenaica*). *Mammalia* 59:229–235.

Ferguson, S. H., A. T. Bergerud, and R. Ferguson. 1988. Predation risk and habitat selection in the persistence of a remnant caribou population. *Oecologia* 76:236–245.

Festa-Bianchet, M. 1986. Seasonal dispersion of overlapping mountain sheep ewe groups. *Journal of Wildlife Management* 50:325–330.

———. 1988a. A pneumonia epizootic in bighorn sheep, with comments on preventive management. *Proceedings of the Biennial Symposium of the Northern Wild Sheep and Goat Council* 6:66–76.

———. 1988b. Age-specific reproduction of bighorn ewes in Alberta, Canada. *Journal of Mammalogy* 69:157–160.

———. 1988c. Birthdate and survival in bighorn lambs (*Ovis canadensis*). *Journal of Zoology* 214:653–661.

———. 1988d. Seasonal range selection in bighorn sheep: conflicts between forage quality, forage quantity, and predator avoidance. *Oecologia* 75:580–586.

———. 1989a. Individual differences, parasites, and the costs of reproduction for bighorn ewes (*Ovis canadensis*). *Journal of Animal Ecology* 58:785–795.

———. 1989b. Survival of male bighorn sheep in southwestern Alberta. *Journal of Wildlife Management* 53:259–263.

———. 1991. The social system of bighorn sheep: grouping patterns, kinship and female dominance rank. *Animal Behaviour* 42:71–82.

————. 1992. Use of age ratios to predict bighorn sheep population dynamics. *Proceedings of the Biennial Symposium of the Northern Wild Sheep and Goat Council* 8:227–236.

————. 1996. Offspring sex ratio studies of mammals: does publication depend upon the quality of the data or the direction of the results? *Ecoscience* 3:42–44.

————. 1998. Condition-dependent reproductive success in bighorn ewes. *Ecology Letters* 1:91–94.

————. 2003. Exploitative wildlife management as a selective pressure for the life-history evolution of large mammals. 191–207 in *Animal behavior and wildlife conservation*, ed. M. Festa-Bianchet, M. Apollonio. Washington, DC: Island Press.

Festa-Bianchet, M., D. W. Coltman, L. Turelli, and J. T. Jorgenson. 2004. Relative allocation to horn and body growth in bighorn rams varies with resource availability. *Behavioral Ecology* 15:305–312.

Festa-Bianchet, M., T. Coulson, J. M. Gaillard, J. T. Hogg, and F. Pelletier. 2006. Stochastic predation and population persistence in bighorn sheep. *Proceedings of the Royal Society of London B* 273:1537–1543.

Festa-Bianchet, M., J.-M. Gaillard, and S. D. Côté. 2003. Variable age structure and apparent density-dependence in survival of adult ungulates. *Journal of Animal Ecology* 72:640–649.

Festa-Bianchet, M., J.-M. Gaillard, and J. T. Jorgenson. 1998. Mass- and density-dependent reproductive success and reproductive costs in a capital breeder. *American Naturalist* 152:367–379.

Festa-Bianchet, M., and J. T. Jorgenson. 1994. Effects of age of primiparity upon horn growth in bighorn ewes. *Proceedings of the Biennial Symposium of the Northern Wild Sheep and Goat Council* 9:116–120.

————. 1998. Selfish mothers: reproductive expenditure and resource availability in bighorn ewes. *Behavioral Ecology* 9:144–150.

Festa-Bianchet, M., J. T. Jorgenson, C. H. Bérubé, C. Portier, and W. D. Wishart. 1997. Body mass and survival of bighorn sheep. *Canadian Journal of Zoology* 75:1372–1379.

Festa-Bianchet, M., J. T. Jorgenson, W. J. King, K. G. Smith, and W. D. Wishart. 1996. The development of sexual dimorphism: seasonal and lifetime mass changes of bighorn sheep. *Canadian Journal of Zoology* 74:330–342.

Festa-Bianchet, M., J. T. Jorgenson, M. Lucherini, and W. D. Wishart. 1995. Life-history consequences of variation in age of primiparity in bighorn ewes. *Ecology* 76:871–881.

Festa-Bianchet, M., J. T. Jorgenson, and D. Réale. 2000. Early development, adult mass, and reproductive success in bighorn sheep. *Behavioral Ecology* 11:633–639.

Festa-Bianchet, M., M. Urquhart, and K. G. Smith. 1994. Mountain goat recruitment: kid production and survival to breeding age. *Canadian Journal of Zoology* 72:22–27.

Fisher, D. O., I. P. F. Owens, and C. N. Johnson. 2001. The ecological basis of life history variation in marsupials. *Ecology* 82:3531–3540.

Foster, B. R., and E. Y. Rahs. 1985. A study of canyon-dwelling mountain goats in relation to proposed hydroelectric development in northwestern British Columbia, Canada. *Biological Conservation* 33:209–228.

Fournier, F., and M. Festa-Bianchet. 1995. Social dominance in adult female mountain goats. *Animal Behaviour* 49:1449–1459.

Fowler, C. W. 1987. A review of density dependence in populations of large mammals. 401–441 in *Current Mammalogy*, ed. H. H. Genoways. New York: Plenum Press.

Fox, J. L., and C. A. Smith. 1988. Winter mountain goat diets in southeast Alaska. *Journal of Wildlife Management* 52:362–365.

Gaillard, J.-M., D. Allainé, D. Pontier, N. G. Yoccoz, and D. E. L. Promislow. 1994. Senescence in natural populations of mammals: a reanalysis. *Evolution* 48:509–516.

Gaillard, J.-M., J.-M. Boutin, D. Delorme, G. V. Laere, P. Duncan, and J.-D. Lebreton. 1997. Early survival in roe deer: causes and consequences of cohort variation in two contrasted populations. *Oecologia* 112:502–513.

Gaillard, J.-M., D. Delorme, J.-M. Boutin, G. V. Laere, and B. Boisaubert. 1996. Body mass of roe deer fawns during winter in 2 contrasting populations. *Journal of Wildlife Management* 60:29–36.

Gaillard, J.-M., D. Delorme, J.-M. Boutin, G. V. Laere, B. Boisaubert, and R. Pradel. 1993a. Roe deer survival patterns: a comparative analysis of contrasting populations. *Journal of Animal Ecology* 62:778–791.

Gaillard, J.-M., D. Delorme, J. M. Jullien, and D. Tatin. 1993b. Timing and synchrony of births in roe deer. *Journal of Mammalogy* 74:738–744.

Gaillard, J.-M., M. Festa-Bianchet, and N. G. Yoccoz. 1998. Population dynamics of large herbivores: variable recruitment with constant adult survival. *Trends in Ecology and Evolution* 13:58–63.

Gaillard, J.-M., M. Festa-Bianchet, N. G. Yoccoz, A. Loison, and C. Toïgo. 2000a. Temporal variation in fitness components and population dynamics of large herbivores. *Annual Review of Ecology and Systematics* 31:367–393.

Gaillard, J.-M., A. Loison, and C. Toïgo. 2003. Variation in life history traits and realistic population models for wildlife management: the case of ungulates. 115–132 in *Animal behavior and wildlife conservation*, ed. M. Festa-Bianchet, M. Apollonio. Washington, DC: Island Press.

Gaillard, J.-M., A. J. Sempéré, J.-M. Boutin, G. V. Laere, and B. Boisaubert. 1992. Effects of age and body weight on the proportion of females breeding in a population of roe deer (*Capreolus capreolus*). *Canadian Journal of Zoology* 70:1541–1545.

Gaillard, J.-M., and N. G. Yoccoz. 2003. Temporal variation in survival of mammals: A case of environmental canalization? *Ecology* 84:3294–3306.

Gaillard, J. M., M. Festa-Bianchet, D. Delorme, and J. Jorgenson. 2000b. Body mass and individual fitness in female ungulates: bigger is not always better. *Proceedings of the Royal Society of London B* 267:471–477.

Gaillard, J. M., M. Festa-Bianchet, and N. G. Yoccoz. 2001. Not all sheep are equal. *Science* 292:1499–1500.

Gallant, B. Y., D. Réale, and M. Festa-Bianchet. 2001. Does mass change of primiparous bighorn ewes reflect reproductive effort? *Canadian Journal of Zoology* 79:312–318.

Garel, M., J.-M. Cugnasse, J.-M. Gaillard, A. Loison, P. Gibert, P. Douvre, and D. Dubray. 2005. Reproductive output of female mouflon (*Ovis gmelini musimon* x *Ovis* sp.): a comparative analysis. *Journal of Zoology* 266:65–71.

Geist, V. 1964. On the rutting behavior of the mountain goat. *Journal of Mammalogy* 45:551–568.

———. 1966. Validity of horn segment counts in aging bighorn sheep. *Journal of Wildlife Management* 30:634–646.

———. 1967. On fighting injuries and dermal shields of mountain goats. *Journal of Wildlife Management* 31:192–194.

———. 1971. *Mountain sheep*. Chicago: University of Chicago Press.

———. 1987. On speciation in ice age mammals, with special reference to cervids and caprids. *Canadian Journal of Zoology* 65:1067–1084.

Gendreau, Y., S. D. Côté, and M. Festa-Bianchet. 2005. Maternal effects on post-weaning physical and social development in juvenile mountain goats (*Oreamnos americanus*). *Behavioral Ecology and Sociobiology* 58:237–246.

Gerard, J.-F., Y. L. Pendu, M.-L. Maublanc, J.-P. Vincent, M.-L. Poulle, and C. Cibien. 1995. Large group formation in European roe deer: an adaptive feature? *Revue d'Écologie (Terre Vie)* 50:391–401.

Gerhart, K. L., D. E. Russell, D. V. DeWetering, R. G. White, and R. D. Cameron. 1997. Pregnancy of adult caribou (*Rangifer tarandus*): evidence for lactational infertility. *Journal of Zoology* 242:17–30.

Giacometti, M., B. Bassano, V. Peracino, and P. Ratti. 1997. Die konstitution des Alpenstein-bockes (*Capra i. ibex* L.) in Abhängigkeit von Geschlecht, Alter, Herkunft und Jahreszeit in Graubünden (Schweiz) und im Parco Nazionale Gran Paradiso (Italien). *Zeitschrift Jagdwissenchaft* 43:24–34.

Glasgow, W. M., T. C. Sorensen, H. D. Carr, and K. G. Smith, 2003. *Management plan for mountain goats in Alberta.* Edmonton: Alberta Fish & Wildlife, 125 pp.

Gomendio, M., T. H. Clutton-Brock, S. D. Albon, F. E. Guinness, and M. J. Simpson. 1990. Mammalian sex ratios and variation in costs of rearing sons and daughters. *Nature* 343:261–263.

Gonzalez-Voyer, A., K. G. Smith, and M. Festa-Bianchet. 2001. Efficiency of aerial censuses of mountain goats. *Wildlife Society Bulletin* 29:140–144.

———. 2003. Dynamics of hunted and unhunted mountain goat populations. *Wildlife Biology* 9:213–218.

Gordon, I. J., A. J. Hester, and M. Festa-Bianchet. 2004. The management of wild large her-bivores to meet economic, conservation and environmental objectives. *Journal of Applied Ecology* 41:1021–1031.

Gosling, L. M. 2002. Adaptive behavior and population viability. In *Animal behavior and wildlife conservation*, ed. M. Festa-Bianchet, M. Apollonio, 13–30. Washington, DC: Is-land Press.

Green, W. C. H. 1990. Reproductive effort and associated costs in bison (*Bison bison*): do older mothers try harder? *Behavioral Ecology* 1:148–160.

Green, W. C. H., J. G. Griswold, and A. Rothstein. 1989. Post-weaning associations among bison mothers and daughters. *Animal Behaviour* 38:847–858.

Green, W. C. H., and A. Rothstein. 1991. Trade-offs between growth and reproduction in fe-male bison. *Oecologia* 86:521–527.

Gregg, M. A., M. Bray, K. M. Kilbride, and M. R. Dunbar. 2001. Birth synchrony and survival of pronghorn fawns. *Journal of Wildlife Management* 65:19–24.

Gross, J. E., P. U. Alkon, and M. W. Demment. 1995. Grouping patterns and spatial segrega-tion by Nubian ibex. *Journal of Arid Environments* 30:423–439.

Gross, J. E., F. J. Singer, and M. E. Moses. 2000. Effects of disease, dispersal, and area on bighorn sheep restoration. *Restoration Ecology* 8:25–37.

Groves, P., and G. F. Shields. 1996. Phylogenetics of the Caprinae based on cytochrome b se-quence. *Molecular Phylogeny and Evolution* 5:467–476.

Guinness, F. E., T. H. Clutton-Brock, and S. D. Albon. 1978. Factors affecting calf mortality in red deer. *Journal of Animal Ecology* 47:812–832.

Hamel, S., S. D. Côté, K. G. Smith, and M. Festa-Bianchet. 2006. Population dynamics and harvest potential of mountain goat herds in Alberta. *Journal of Wildlife Management* 70:1044–1053.

Hanski, I. 1999. *Metapopulation ecology*. Oxford: Oxford University Press.

Hare, J. F., and J. O. Murie. 1992. Manipulation of litter size reveals no cost of reproduction in Columbian ground squirrels. *Journal of Mammalogy* 73:449–454.

Harris, R. B., W. A. Wall, and F. W. Allendorf. 2002. Genetic consequences of hunting: what do we know and what should we do? *Wildlife Society Bulletin* 30:634–643.

Hartl, G. B., H. Burger, R. Willing, and F. Suchentrunk. 1990. On the biochemical systemat-ics of the caprini and the rupicaprini. *Biochemical Systematic Ecology* 18:175–182.

Hassanin, A., E. Pasquet, and J.-D. Vigne. 1998. Molecular systematics of the subfamily Caprinae (Artiodactyla, Bovidae) as determined from cytochrome b sequence. *Journal of Mammalian Evolution* 5:217–236.

Haviernick, M. 1996. *La stratégie alimentaire de la chèvre de montagne (*Oreamnos americanus*): étude de l'utilisation de l'habitat et du comportement anti-prédateur.* M.Sc. Thesis, Université de Sherbrooke.

Haviernick, M., S. D. Côté, and M. Festa-Bianchet. 1998. Immobilization of mountain goats with xylazine and reversal with idazoxan. *Journal of Wildlife Diseases* 34:342–347.

Hebert, D. M., and T. Smith. 1986. Mountain goat management in British Columbia. *Proceedings of the Biennial Symposium of the Northern Wild Sheep and Goat Council* 5:48–59.

Hewison, A. J. M., and J.-M. Gaillard. 1996. Birth-sex ratios and local resource competition in roe deer, *Capreolus capreolus. Behavioral Ecology* 7:461–464.

———. 1999. Successful sons or advantaged daughters? The Trivers-Willard model and sex-biased maternal investment in ungulates. *Trends in Ecology and Evolution* 14:229–234.

Hewison, A. J. M., J.-M. Gaillard, M. Festa-Bianchet, and P. Blanchard. 2002. Maternal age is not a predominant determinant of progeny sex ratio variation in ungulates. *Oikos* 98:334–339.

Hjeljord, O., and T. Histøl. 1999. Range-body mass interactions of a northern ungulate—a test of hypothesis. *Oecologia* 119:326–339.

Hobbs, N. T. 1987. Fecal indices to dietary quality: a critique. *Journal of Wildlife Management* 51:317–320.

Hobbs, N. T., M. W. Miller, J. A. Bailey, D. F. Reed, and R. B. Gill. 1990. Biological criteria for introductions of large mammals: using simulation models to predict impacts of competition. *Transactions North American Wildlife and Natural Resources Conference* 55:620–632.

Hoefs, M. 1991. A longevity record for Dall sheep, *Ovis dalli dalli*, Yukon Territory. *Canadian Field-Naturalist* 105:397–398.

Hoefs, M., and I. M. Cowan. 1975. Phytosociological analysis and synthesis of Sheep Mountain, southwest Yukon Territory, Canada. *Syesis (Suppl. 1)* 8:125–228.

Hoefs, M., and R. König. 1984. Reliability of aging old Dall sheep ewes by the horn annulus technique. *Journal of Wildlife Management* 48:980–982.

Hoefs, M., and U. Nowlan. 1997. Comparison of horn growth in captive and free-ranging Dall's rams. *Journal of Wildlife Management* 61:1154–1160.

Hogg, J. T. 1987. Intrasexual competition and mate choice in Rocky Mountain bighorn sheep. *Ethology* 75:119–144.

———. 2000. Mating systems and conservation at large spatial scales. 214–252 in *Vertebrate mating systems*, ed. M. Apollonio, M. Festa-Bianchet, D. Mainardi. Singapore: World Scientific.

Hogg, J. T., and S. H. Forbes. 1997. Mating in bighorn sheep: frequent male reproduction via a high-risk "unconventional" tactic. *Behavioral Ecology and Sociobiology* 41:33–48.

Hogg, J. T., C. C. Hass, and D. A. Jenni. 1992. Sex-biased maternal expenditure in Rocky Mountain bighorn sheep. *Behavioral Ecology and Sociobiology* 31:243–251.

Holand, Ø., K. H. Roed, A. Mysterud, J. Kumpula, M. Nieminen, and M. E. Smith. 2003. The effect of sex ratio and male age structure on reindeer calving. *Journal of Wildlife Management* 67:25–33.

Houle, D. 1991. Genetic covariance of fitness correlates: what genetic correlations are made of and why it matters. *Evolution* 45:630–648.

Houston, D. B. 1995. Response to inaccurate data and the Olympic National Park mountain goat controversy. *Northwest Science* 69:239–240.

Houston, D. B., C. T. Robbins, and V. Stevens. 1989. Growth in wild and captive mountain goats. *Journal of Mammalogy* 70:412–416.

Houston, D. B., and E. G. Schreiner. 1995. Alien species in national parks: drawing lines in space and time. *Conservation Biology* 9:204–209.

Houston, D. B., and V. Stevens. 1988. Resource limitation in mountain goats: a test by experimental cropping. *Canadian Journal of Zoology* 66:228–238.

Hughes, L. 2000. Biological consequences of global warming: is the signal already apparent? *Trends in Ecology and Evolution* 15:56–61.

Humphries, M. M., and S. Boutin. 1996. Reproductive demands and mass gains: a paradox in female red squirrels (*Tamiasciurus hudsonicus*). *Journal of Animal Ecology* 65:332–338.

Hutchings, J. A. 1993. Adaptive life histories affected by age-specific survival and growth rate. *Ecology* 74:673–684.

Hutchings, M. 1995. Olympic mountain goat controversy continues. *Conservation Biology* 9:1324–1326.

Hutchins, M. 1984. *The mother–offspring relationship in mountain goats* (Oreamnos americanus). Ph.D. thesis, University of Washington.

Iason, G. R., and F. E. Guinness. 1985. Synchrony of oestrus and conception in red deer (*Cervus elaphus* L.). *Journal of Animal Ecology* 33:1169–1174.

Illius, A. W., and C. Fitzgibbon. 1994. Costs of vigilance in foraging ungulates. *Animal Behaviour* 47:481–484.

Iriarte, J. A., W. L. Franklin, W. E. Johnson, and K. H. Redford. 1990. Biogeographic variation of food habits and body size of the American puma. *Oecologia* 85:185–190.

Jacobson, A. R., A. Provenzale, A. v. Hardenberg, B. Bassano, and M. Festa-Bianchet. 2004. Climate forcing and density-dependence in a mountain ungulate population. *Ecology* 85:1598–1610.

James, A. R. C., and A. K. Stuart-Smith. 2000. Distribution of caribou and wolves in relation to linear corridors. *Journal of Wildlife Management* 64:154–159.

Jarman, P. J. 1974. The social organisation of antelope in relation to their ecology. *Behaviour* 48:215–267.

Jass, C. N., J. I. Mead, and L. E. Logan. 2000. Harrington's extinct mountain goat (*Oreamnos harringtoni* Stock 1936) from Muskox Cave, New Mexico. *Texas Journal of Science* 52:121–132.

Jedrzejewski, W., B. Jedrzejewska, H. Okarma, K. Schmidt, K. Zub, and M. Musiani. 2000. Prey selection and predation by wolves in Bialowieza Primeval Forest, Poland. *Journal of Mammalogy* 81:197–212.

Johnson, C. N. 1986a. Philopatry, reproductive success of females, and maternal investment in the red-necked wallaby. *Behavioral Ecology and Sociobiology* 19:143–150.

Johnson, R. L. 1986b. Mountain goat management in Washington. *Proceedings of the Biennial Symposium of the Northern Wild Sheep and Goat Council* 5:60–62.

Jorgenson, J. T., M. Festa-Bianchet, J.-M. Gaillard, and W. D. Wishart. 1997. Effects of age, sex, disease, and density on survival of bighorn sheep. *Ecology* 78:1019–1032.

Jorgenson, J. T., M. Festa-Bianchet, M. Lucherini, and W. D. Wishart. 1993a. Effects of body size, population density and maternal characteristics on age of first reproduction in bighorn ewes. *Canadian Journal of Zoology* 71:2509–2517.

Jorgenson, J. T., M. Festa-Bianchet, and W. D. Wishart. 1993b. Harvesting bighorn ewes: consequences for population size and trophy ram production. *Journal of Wildlife Management* 57:429–435.

———. 1998. Effects of population density on horn development in bighorn rams. *Journal of Wildlife Management* 62:1011–1020.

Jorgenson, J. T., J. Samson, and M. Festa-Bianchet. 1990. Field immobilization of bighorn sheep with xylazine hydrochloride and antagonism with idazoxan. *Journal of Wildlife Diseases* 26:522–527.

Joslin, G. 1986. Mountain goat population changes in relation to energy exploration along Montana's Rocky Mountain front. *Proceedings of the Biennial Symposium of the Northern Wild Sheep and Goat Council* 5:253–271.

Käär, P., and J. Jokela. 1998. Natural selection on age-specific fertilities in human females: comparison of individual-level fitness measures. *Proceedings of the Royal Society of London B* 265:2415–2420.

Kamler, J., and M. Homolka. 2005. Faecal nitrogen: a potential indicator of red and roe deer diet quality in forest habitats. *Folia Zoologica* 54:89–98.

Karubian, J., and J. P. Swaddle. 2001. Selection on females can create "larger males." *Proceedings of the Royal Society of London B* 1468:725–728.

Keech, M. A., R. T. Bowyer, J. M. Ver Hoef, R. D. Boertje, B. W. Dale, and T. R. Stephenson. 2000. Life-history consequences of maternal condition in Alaskan moose. *Journal of Wildlife Management* 64:450–462.

Keller, L. F., and D. M. Waller. 2002. Inbreeding effects in wild populations. *Trends in Ecology and Evolution* 17:230–241.

King, W. J., and J. O. Murie. 1985. Temporal overlap of female kin in Columbian ground squirrels (*Spermophilus columbianus*). *Behavioral Ecology and Sociobiology* 16:337–341.

Kinley, T. A., and C. D. Apps. 2001. Mortality patterns in a subpopulation of endangered mountain caribou. *Wildlife Society Bulletin* 29:158–164.

Kojola, I., and E. Eloranta. 1989. Influences of maternal body weight, age and parity on sex ratio in semidomesticated reindeer (*Rangifer t. tarandus*). *Evolution* 43:1331–1336.

Kojola, I., and T. Helle. 1993. Calf harvest and reproductive rate of reindeer in Finland. *Journal of Wildlife Management* 57:451–453.

Kokko, H. 2001. Optimal and suboptimal use of compensatory response to harvesting: timing of hunting as an example. *Wildlife Biology* 7:141–149.

Kruuk, L. E. B., T. H. Clutton-Brock, S. D. Albon, J. M. Pemberton, and F. E. Guinness. 1999a. Population density affects sex ratio variation in red deer. *Nature* 399:459–461.

Kruuk, L. E. B., T. H. Clutton-Brock, K. E. Rose, and F. E. Guinness. 1999b. Early determinants of lifetime reproductive success differ between the sexes in red deer. *Proceedings of the Royal Society of London B* 266:1655–1661.

Kruuk, L. E. B., T. H. Clutton-Brock, J. Slate, J. M. Pemberton, S. Brotherstone, and F. E. Guinness. 2000. Heritability of fitness in a wild mammal population. *Proceedings of the National Academy of Sciences USA* 97:698–703.

Kruuk, L. E. B., J. Slate, J. M. Pemberton, S. Brotherstone, F. Guinness, and T. Clutton-Brock. 2002. Antler size in red deer: heritability and selection but no evolution. *Evolution* 56:1683–1695.

Kuck, L. 1977. The impact of hunting on Idaho's Pahsimeroi mountain goat herd. *International Mountain Goat Symposium* 1:114–125.

Kunkel, K., and D. H. Pletscher. 1999. Species-specific population dynamics of cervids in a multipredator ecosystem. *Journal of Wildlife Management* 63:1082–1093.

L'Heureux, N., M. Lucherini, M. Festa-Bianchet, and J. T. Jorgenson. 1995. Density-dependent mother–yearling association in bighorn sheep. *Animal Behaviour* 49:901–910.

Langvatn, R., S. D. Albon, T. Burkey, and T. H. Clutton-Brock. 1996. Climate, plant phenology and variation in age of first reproduction in a temperate herbivore. *Journal of Animal Ecology* 65:653–670.

Langvatn, R., and A. Loison. 1999. Consequences of harvesting on age structure, sex ratio and population dynamics of red deer *Cervus elaphus* in central Norway. *Wildlife Biology* 5:213–223.

Larsen, D. G., D. A. Gauthier, and R. L. Markel. 1989. Causes and rate of moose mortality in the southwest Yukon. *Journal of Wildlife Management* 53:548–557.

Laundré, J. W. 1994. Resource overlap between mountain goats and bighorn sheep. *Great Basin Naturalist* 54:114–121.

Laurance, W. F. 2001. Future shock: forecasting a grim fate for the Earth. *Trends in Ecology and Evolution* 16:531–533.

Lawrence, A. B. 1990. Mother–daughter and peer relationships of Scottish hill sheep. *Animal Behaviour* 39:481–486.

Lawrence, A. B., and D. G. M. Wood-Gush. 1988. Home-range behaviour and social organization of Scottish blackface sheep. *Journal of Applied Ecology* 25:25–40.

Leberg, P. L., and M. H. Smith. 1993. Influence of density on growth of white-tailed deer. *Journal of Mammalogy* 74:723–731.

Leblanc, M., M. Festa-Bianchet, and J. T. Jorgenson. 2001. Sexual size dimorphism in bighorn sheep (*Ovis canadensis*): effects of population density. *Canadian Journal of Zoology* 79:1661–1670.

Lebreton, J.-D. 2005. Dynamical and statistical models for exploited populations. *Australia and New Zealand Journal of Statistics* 47:49–63.

Lebreton, J.-D., K. P. Burnham, J. Clobert, and D. R. Anderson. 1992. Modeling survival and testing biological hypotheses using marked animals: a unified approach with case studies. *Ecological Monographs* 62:67–118.

Lemke, T. O. 2004. Origin, expansion, and status of mountain goats in Yellowstone National Park. *Wildlife Society Bulletin* 32:532–541.

Lentfer, J. W. 1955. A two-year study of the Rocky Mountain goat in the Crazy Mountains, Montana. *Journal of Wildlife Management* 19:417–429.

Lesage, L., M. Crête, J. Huot, and J. P. Ouellet. 2001. Evidence for a trade-off between growth and body reserves in northern white-tailed deer. *Oecologia* 126:30–41.

Leslie, D. M., and E. E. Starkey. 1987. Fecal indices to dietary quality: a reply. *Journal of Wildlife Management* 51:321–325.

Linklater, W. L. 2000. Adaptive explanation in socio-ecology: lessons from the Equidae. *Biological Review of the Cambridge Philosophical Society* 75:1–20.

Linnell, J. D. C., R. Aanes, and R. Andersen. 1995. Who killed Bambi? The role of predation in the neonatal mortality of temperate ungulates. *Wildlife Biology* 1:209–223.

Locati, M. 1990. Female chamois defends kids from eagle attack. *Mammalia* 54:155–156.

Locati, M., and S. Lovari. 1990. Sexual differences in aggressive behaviour of the Apennine chamois. *Ethology* 84:295–306.

Loison, A., M. Festa-Bianchet, J.-M. Gaillard, J. T. Jorgenson, and J.-M. Jullien. 1999a. Age-specific survival in five populations of ungulates: evidence of senescence. *Ecology* 80:2539–2554.

Loison, A., J.-M. Gaillard, and H. Houssin. 1994. New insight on survivorship of female chamois (*Rupicapra rupicapra*) from marked animals. *Canadian Journal of Zoology* 72:591–597.

Loison, A., J.-M. Gaillard, and J.-M. Jullien. 1996. Demographic patterns after an epizootic of keratoconjunctivitis in a chamois population. *Journal of Wildlife Management* 60:517–527.

Loison, A., J.-M. Gaillard, C. Pélabon, and N. G. Yoccoz. 1999b. What factors shape sexual size dimorphism in ungulates? *Evolution and Ecology Research* 1:611–633.

Loison, A., and R. Langvatn. 1998. Short- and long-term effects of winter and spring weather on growth and survival of red deer in Norway. *Oecologia* 116:489–500.

Lunn, N. J., I. L. Boyd, and J. P. Croxall. 1994. Reproductive performance of female Antarctic fur seals: the influence of age, breeding experience, environmental variation and individual quality. *Journal of Animal Ecology* 63:827–840.

Lyman, R. L. 1988. Significance for wildlife management of the late Quaternary biogeography of mountain goats (*Oreamnos americanus*) in the Pacific Northwest, U.S.A. *Arctic Alpine Research* 20:13–23.

———. 1994. The Olympic mountain goat controversy: a different perspective. *Conservation Biology* 8:898–901.

———. 1995. Inaccurate data and the Olympic National Park mountain goat controversy. *Northwest Science* 69:234–238.

———. 1999. Politics, book reviews, science, and white lies: a response to R. Geral Wright. *Northwest Science* 73:131–134.

Lynch, G. M., B. Lajeunesse, J. Willman, and E. S. Telfer. 1995. Moose weights and measurements from Elk Island National Park. *Alces* 31:199–207.

MacCallum, B. N., and V. Geist. 1992. Mountain restoration: soil and surface wildlife habitat. *GeoJournal* 27:23–46.

MacWhirter, R. B. 1991. Effects of reproduction on activity and foraging behaviour of adult female Columbian ground squirrels. *Canadian Journal of Zoology* 69:2209–2216.

Main, M. B., F. W. Weckerly, and V. C. Bleich. 1996. Sexual segregation in ungulates: new directions for research. *Journal of Mammalogy* 77:449–461.

Marshall, T. C., D. W. Coltman, J. M. Pemberton, J. Slate, J. A. Spalton, F. E. Guinness, J. A. Smith, J. G. Pilkington, and T. H. Clutton-Brock. 2002. Estimating the prevalence of inbreeding from incomplete pedigrees. *Proceedings of the Royal Society of London B* 269:1533–1539.

Martinez, M., C. Rodrıguez-Vigal, O. R. Jones, T. Coulson, and A. S. Miguel. 2005. Different hunting strategies select for different weights in red deer. *Biology Letters* 1:353–356.

McCullough, D. R. 1994. What do herd composition counts tell us? *Wildlife Society Bulletin* 22:295–300.

McElligott, A. G., R. Altwegg, and T. J. Hayden. 2002. Age-specific survival and reproductive probabilities: evidence for senescence in male fallow deer (*Dama dama*). *Proceedings of the Royal Society of London B* 269:1129–1137.

McElligott, A. G., M. P. Gammell, H. C. Harty, D. R. Paini, D. T. Murphy, J. T. Walsh, and T. J. Hayden. 2001. Sexual size dimorphism in fallow deer (*Dama dama*): do larger, heavier males gain greater mating success? *Behavioral Ecology and Sociobiology* 49:266–272.

McElligott, A. G., F. Naulty, W. V. Clarke, and T. J. Hayden. 2003. The somatic cost of reproduction: what determines reproductive effort in prime-aged fallow bucks? *Evolution and Ecology Research* 5:1239–1250.

McFetridge, R. J. 1977. *Strategy of resource use by mountain goats in Alberta.* M.Sc. thesis, University of Alberta, Edmonton.

McGraw, J. B., and H. Caswell. 1996. Estimation of individual fitness from life-history data. *American Naturalist* 147:47–64.

McNamara, J. M., and A. I. Houston. 1996. State-dependent life histories. *Nature* 380:215–221.

Messier, F. 1991. The significance of limiting and regulating factors on the demography of moose and white-tailed deer. *Journal of Animal Ecology* 60:377–393.

———. 1994. Ungulate population models with predation: a case study with the North American moose. *Ecology* 75:478–488.

Messier, F., and D. O. Joly. 2000. Comment: regulation of moose populations by wolf predation. *Canadian Journal of Zoology* 78:506–510.

Miller, F. L., and A. Gunn. 1979. Responses of Peary caribou and muskoxen to helicopter harassment. *Canadian Wildlife Service Occasional Paper* 40:1–90.

Milner, J. M., C. Bonenfant, A. Mysterud, J. M. Gaillard, S. Csany, and N. C. Stenseth. 2006. Temporal and spatial development of red deer harvesting in Europe: biological and cultural factors. *Journal of Applied Ecology* 43:721–734.

Miquelle, D. G. 1990. Why don't bull moose eat during the rut? *Behavioral Ecology and Sociobiology* 27:145–151.

Miura, S. 1986. Body and horn growth patterns in the Japanese serow, *Capricornis crispus*. *Journal of Mammalogical Society of Japan* 11:1–13.

Miura, S., I. Kita, and M. Sugimura. 1987. Horn growth and reproductive history in female Japanese serow. *Journal of Mammalogy* 68:826–836.

Miura, S., and K. Yasui. 1985. Validity of tooth eruption-wear patterns as age criteria in the Japanese serow, *Capricornis crispus*. *Journal of Mammalogy Society of Japan* 10:169–178.

Monard, A.-M., P. Duncan, H. Fritz, and C. Feh. 1997. Variations in the birth sex ratio and neonatal mortality in a natural herd of horses. *Behavioral Ecology and Sociobiology* 41:243–249.

Mooring, M. S., T. A. Fitzpatrick, J. E. Benjamin, I. C. Fraser, T. T. Nishihira, D. D. Reisig, and E. M. Rominger. 2003. Sexual segregation in desert bighorn sheep (*Ovis canadensis Mexicana*). *Behaviour* 140:183–207.

Moreno, J., J. Potti, and S. Merino. 1997. Parental energy expenditure and offspring size in the pied flycatcher *Ficedula hypoleuca*. *Oikos* 79:559–567.

Mysterud, A., E. Meisingset, R. Langvatn, N. G. Yoccoz, and N. C. Stenseth. 2005a. Climate-dependent allocation of resources to secondary sexual traits in red deer. *Oikos* 111:245–252.

Mysterud, A., F. J. Pérez-Barbería, and I. J. Gordon. 2001. The effect of season, sex and feeding style on home range area versus body mass scaling in temperate ruminants. *Oecologia* 127:30–39.

Mysterud, A., E. J. Solberg, and N. G. Yoccoz. 2005b. Aging and reproductive effort in male moose under variable levels of intrasexual competition. *Journal of Animal Ecology* 74:742–754.

Mysterud, A., G. Steinheim, N. G. Yoccoz, Ø. Holand, and N. C. Stenseth. 2002. Early onset of reproductive senescence in domestic sheep, *Ovis aries*. *Oikos* 97:177–183.

Mysterud, A., P. Trjanowski, and M. Panek. 2006. Selectivity of harvesting differs between local and foreign roe deer hunters: trophy stalkers have the first shot at the right time. *Biology Letters* 2:632–635.

Mysterud, A., N. G. Yoccoz, N. C. Stenseth, and R. Langvatn. 2000. Relationships between sex ratio, climate and density in red deer: the importance of spatial scale. *Journal of Animal Ecology* 69:959–974.

Nagorsen, D. W., and G. Keddie. 2000. Late Pleistocene mountain goats (*Oreamnos americanus*) from Vancouver Island: biogeographic implications. *Journal of Mammalogy* 81:666–675.

Nelson, M. E. 1998. Development of migratory behavior in northern white-tailed deer. *Canadian Journal of Zoology* 76:426–432.

Nelson, M. E., and L. D. Mech. 1987. Demes within a northeastern Minnesota deer population. 27–40 in *Mammalian dispersal patterns*, ed. B. D. Chepko-Sade, Z. T. Halpin. Chicago: University of Chicago Press.

Nixon, C. M., L. P. Hansen, P. A. Brewer, J. E. Chelsvig, T. L. Esker, D. Etter, J. B. Sullivan, R. G. Koerkenmeier, and P. C. Mankin. 2001. Survival of white-tailed deer in intensively farmed areas of Illinois. *Canadian Journal of Zoology* 79:581–588.

Noordwijk, A. J. v., and G. de Jong. 1986. Acquisition and allocation of resources: their influence on variation in life history tactics. *American Naturalist* 128:137–142.

Nygrén, T., and M. Pesonen. 1993. The moose population (*Alces alces* L.) and methods of moose management in Finland, 1975–89. *Finnish Game Research* 48:46–53.

Ochiai, K., and K. Susaki. 2002. Effects of territoriality on population density in the Japanese serow (*Capricornis crispus*). *Journal of Mammalogy* 83:964–972.

Ochiai, K., and K. Susaki. 2007. Causes of natal dispersal in a monogamous ungulate, the Japanese serow, *Capricornis crispus*. *Animal Behaviour* 73:125–131.

Owen-Smith, N. 1990. Demography of a large herbivore, the greater kudu *Tragelaphus strepsiceros*, in relation to rainfall. *Journal of Animal Ecology* 59:893–913.

Owen-Smith, N., D. R. Mason, and J. O. Ogutu. 2005. Correlates of survival rates for 10 African ungulate populations: density, rainfall and predation. *Journal of Animal Ecology* 74:774–788.

Ozoga, J. J., and L. J. Verme. 1984. Effect of family-bond deprivation on reproductive performance in female white-tailed deer. *Journal of Wildlife Management* 48:1326–1334.

Packer, C. 1983. Sexual dimorphism: the horns of African antelopes. *Science* 221:1191–1193.

Palmer, A. R. 1999. Detecting publication bias in meta-analyses: a case study of fluctuating asymmetry and sexual selection. *American Naturalist* 154:220–233.

———. 2000. Quasireplication and the contract of error: lessons from sex ratios, heritabilities and fluctuating asymmetry. *Annual Review of Ecology and Systematics* 31:441–480.

Partridge, L. 1992. Measuring reproductive costs. *Trends in Ecology and Evolution* 7:99–100.

Partridge, L., and M. Mangel. 1999. Messages from mortality: the evolution of death rates in the old. *Trends in Ecology and Evolution* 14:438–442.

Pelletier, F. 2005. Foraging time of rutting bighorn rams varies with individual behavior, not mating tactic. *Behavioral Ecology* 16:280–285.

Pelletier, F., and M. Festa-Bianchet. 2004. Effects of body mass, age, dominance and parasite load on foraging time of bighorn rams, *Ovis canadensis*. *Behavioral Ecology and Sociobiology* 56:546–551.

Pelletier, F., J. T. Hogg, and M. Festa-Bianchet. 2004. Effect of chemical immobilization on social status of bighorn rams. *Animal Behaviour* 67:1163–1165.

Pemberton, J. M., S. D. Albon, F. E. Guinness, T. H. Clutton-Brock, and G. A. Dover. 1992. Behavioral estimates of male mating success tested by DNA fingerprinting in a polygynous mammal. *Behavioral Ecology* 3:66–75.

Penner, D. F. 1988. Behavioral response and habituation of mountain goats in relation to petroleum exploration at Pinto Creek, Alberta. *Proceedings of the Biennial Symposium of the Northern Wild Sheep and Goat Council* 6:141–158.

Pérez-Barbería, F. J., L. Robles, and C. Nores. 1996. Horn growth pattern in Cantabrian chamois *Rupicapra pyrenaica parva*: influence of sex, location and phaenology. *Acta Theriologica* 41:83–92.

Peters, R. H. 1983. *The ecological implications of body size*. Cambridge: Cambridge University Press.

Peterson, R. O., N. J. Thomas, J. M. Thurber, J. A. Vucetich, and T. A. Waite. 1998. Population limitation and the wolves of Isle Royale. *Journal of Mammalogy* 79:828–841.

Pettifor, R. A. 1993. Brood-manipulation experiments, II: A cost of reproduction in blue tits (*Parus caeruleus*)? *Journal of Animal Ecology* 62:145–159.

Pettorelli, N., S. D. Côté, A. Gingras, F. Potvin, and J. Huot. 2007. Aerial surveys vs hunting statistics to monitor deer density: the example of Anticosti Island (Québec, Canada). *Wildlife Biology* 13:in press.

Pettorelli, N., J. M. Gaillard, G. Van Laere, P. Duncan, P. Kjellander, O. Liberg, D. Delorme, and D. Maillard. 2002. Variations in adult body mass in roe deer: the effects of population density at birth and of habitat quality. *Proceedings of the Royal Society of London B* 269:747–753.

Pfitsch, W. A., and L. C. Bliss. 1985. Seasonal forage availability and potential vegetation limitations to a mountain goat population, Olympic National Park. *American Midland Naturalist* 113:109–121.

Picard, K., D. W. Thomas, M. Festa-Bianchet, and C. Lanthier. 1994. Bovid horns: an important site for heat loss during winter? *Journal of Mammalogy* 75:710–713.

Portier, C., M. Festa-Bianchet, J.-M. Gaillard, J. T. Jorgenson, and N. G. Yoccoz. 1998. Effects of density and weather on survival of bighorn sheep lambs (*Ovis canadensis*). *Journal of Zoology* 245:271–278.

Post, E., M. C. Forchhammer, N. C. Stenseth, and R. Langvatn. 1999a. Extrinsic modification of vertebrate sex ratios by climatic variation. *American Naturalist* 154:194–204.

Post, E., R. O. Peterson, N. C. Stenseth, and B. E. McLaren. 1999b. Ecosystem consequences of wolf behavioural response to climate. *Nature* 401:905–907.

Rachlow, J. L., and R. T. Bowyer. 1991. Interannual variation in timing and synchrony of parturition in Dall's sheep. *Journal of Mammalogy* 72:487–492.

Réale, D., and M. Festa-Bianchet. 2000. Mass-dependent reproductive strategies in wild bighorn ewes: a quantitative genetic approach. *Journal of Evolutionary Biology* 13:679–688.

Réale, D., B. Y. Gallant, M. Leblanc, and M. Festa-Bianchet. 2000. Consistency of temperament in bighorn ewes and correlates with behaviour and life history. *Animal Behaviour*
60:589–597.

Reznick, D. 1985. Costs of reproduction: an evaluation of the empirical evidence. *Oikos*
44:257–267.

———. 1992. Measuring the costs of reproduction. *Trends in Ecology and Evolution* 7:42–45.

Reznick, D., L. Nunnev, and A. Tessier. 2000. Big houses, big cars, superfleas and the costs of
reproduction. *Trends in Ecology and Evolution* 15:421–425.

Rideout, C. B. 1974. *A radio telemetry study of the ecology and behavior of the mountain goat in
western Montana*. Ph.D. thesis, University of Kansas, Lawrence.

———. 1978. Mountain goat. 149–159 in *Big game of North America*, ed. J. L. Schmidt, D. L.
Gilbert. Harrisburg, PA: Stackpole Books.

Rideout, C. B., and R. S. Hoffmann. 1975. *Oreamnos americanus*. *Mammalian Species* 63:1–6.

Risenhoover, K. L., and J. A. Bailey. 1985a. Foraging ecology of mountain sheep: implications
for habitat management. *Journal of Wildlife Management* 49:797–804.

———. 1985b. Relationships between group size, feeding time, and agonistic behavior of
mountain goats. *Canadian Journal of Zoology* 63:2501–2506.

Robbins, C. T., A. E. Hagerman, P. J. Austin, C. McArthur, and T. A. Hanley. 1991. Variation
in mammalian physiological responses to a condensed tannin and its ecological implications. *Journal of Mammalogy* 72:480–486.

Robbins, C. T., T. A. Hanley, A. E. Hagerman, O. Hjeljord, D. L. Baker, C. C. Schwartz, and
W. W. Mautz. 1987. Role of tannins in defending plants against ruminants: reduction in
protein availability. *Ecology* 68:98–107.

Romeo, G., S. Lovari, and M. Festa-Bianchet. 1997. Group leaving in mountain goats: are
young males ousted by adult females? *Behavioral Processes* 40:243–246.

Rose, K. E., T. H. Clutton-Brock, and F. E. Guinness. 1998. Cohort variation in male survival
and lifetime breeding success in red deer. *Journal of Animal Ecology* 67:979–986.

Ross, I. P., M. G. Jalkotzy, and P.-Y. Daoust. 1995. Fatal trauma sustained by cougars, *Felis
concolor*, while attacking prey in southern Alberta. *Canadian Field-Naturalist* 109:261–263.

Ross, P. I., and M. G. Jalkotzy. 1996. Cougar predation on moose in southwestern Alberta. *Alces* 32:1–8.

Ross, P. I., M. G. Jalkotzy, and M. Festa-Bianchet. 1997. Cougar predation on bighorn sheep
in southwestern Alberta during winter. *Canadian Journal of Zoology* 75:771–775.

Ruckstuhl, K. E. 1998. Foraging behaviour and sexual segregation in bighorn sheep. *Animal
Behaviour* 56:99–106.

———. 1999. To synchronise or not to synchronise: a dilemma for young bighorn males? *Behaviour* 136:805–818.

Ruckstuhl, K. E., and M. Festa-Bianchet. 2001. Group choice by subadult bighorn rams:
trade-offs between foraging efficiency and predator avoidance. *Ethology* 107:161–172.

Ruckstuhl, K. E., and P. Neuhaus. 2000. Sexual segregation in ungulates: a new approach. *Behaviour* 137:361–377.

———. 2001. Behavioral synchrony in ibex groups: effects of age, sex and habitat. *Behaviour*
138:1033–1046.

———. 2002. Sexual segregation in ungulates: a comparative test of three hypotheses. *Biological Review* 77:77–96.

Rutberg, A. T. 1986. Lactation and fetal sex ratios in American bison. *American Naturalist*
127:89–94.

Sadleir, R. M. F. S. 1987. Reproduction of female cervids. In *Biology and Management of the
Cervidae*, ed. C. M. Wemmer. Washington, DC: Smithsonian Institution Press.

Sæther, B.-E. 1997. Environmental stochasticity and population dynamics of large herbivores: a search for mechanisms. *Trends in Ecology and Evolution* 12:143–149.

Sæther, B.-E., and M. Heim. 1993. Ecological correlates of individual variation in age at maturity in female moose (*Alces alces*): the effects of environmental variability. *Journal of Animal Ecology* 62:482–489.

Sæther, B. E., M. Lillegard, V. Grøtan, F. Filli, and S. Engen. 2007. Predicting fluctuations of reintroduced ibex populations: the importance of density dependence, environmental stochasticity and uncertain population estimates. *Journal of Animal Ecology* 76:326–336.

Sæther, B. E., E. J. Solberg, and M. Heim. 2003. Effects of altering sex ratio structure on the demography of an isolated moose population. *Journal of Wildlife Management* 67:455–466.

Saltz, D., and B. P. Kotler. 2003. Maternal age is a predominant determinant of progeny sex ratio variation in ungulates: a reply to Hewison et al. *Oikos* 101:646–648.

San José, C., S. Lovari, and N. Ferrari. 1997. Grouping in roe deer: an effect of habitat openness or cover distribution? *Acta Theriologica* 42:235–239.

Schaefer, J. A. 2003. Long-term range recession and the persistence of caribou in the taiga. *Conservation Biology* 17:1435–1439.

Scheffer, V. B. 1993. The Olympic mountain goat controversy: a perspective. *Conservation Biology* 7:916–919.

Shackleton, D. M. 1997. *Wild sheep and goats and their relatives: status survey and conservation action plan for Caprinae.* Gland, Switzerland: IUCN.

Shaw, J. H., and T. S. Carter. 1989. Calving patterns among American Bison. *Journal of Wildlife Management* 53:896–898.

Sheldon, B. C., and S. A. West. 2004. Maternal dominance, maternal condition, and offspring sex ratio in ungulate mammals. *American Naturalist* 163:40–54.

Silk, J. B. 1983. Local resource competition and facultative adjustment of sex ratios in relation to competitive activities. *American Naturalist* 130:56–66.

Sinclair, A. R. E. 1977. *The African buffalo.* Chicago: University of Chicago Press.

———. 1991. Science and the practice of wildlife management. *Journal of Wildlife Management* 55:767–773.

Sinclair, A. R. E., S. Mduma, and J. S. Brashares. 2003. Patterns of predation in a diverse predator–prey system. *Nature* 425:288–290.

Sinclair, A. R. E., and R. P. Pech. 1996. Density dependence, stochasticity, compensation and predator regulation. *Oikos* 75:164–173.

Singer, F. J., and J. L. Doherty. 1985. Movements and habitat use in an unhunted population of mountain goats, *Oreamnos americanus. Canadian Field-Naturalist* 99:205–217.

Skogland, T. 1991. What are the effects of predators on large ungulate populations? *Oikos* 61:401–411.

Smith, B. L. 1976. *Ecology of the Rocky Mountain goat in the Bitterroot Mountains, Montana.* M.Sc. thesis, University of Montana, Missoula.

———. 1988a. Criteria for determining age and sex of American mountain goats in the field. *Journal of Mammalogy* 69:395–402.

Smith, B. L., and S. H. Anderson. 1998. Juvenile survival and population regulation of the Jackson elk herd. *Journal of Wildlife Management* 62:1036–1045.

Smith, C. A. 1986. Rates and causes of mortality in mountain goats in southeast Alaska. *Journal of Wildlife Management* 50:743–746.

Smith, K. G. 1988b. Factors affecting the population dynamics of mountain goats in west-central Alberta. Proceedings of the Biennial Symposium of the Northern Wild Sheep and Goat Council 6:308–329.

———. 2004. *Woodland caribou demography and persistence relative to landscape change in west-central Alberta.* M.Sc. thesis, University of Alberta, Edmonton.

Smith, K. G., E. J. Ficht, D. Hobson, T. C. Sorensen, and D. Hervieux. 2000. Winter distribution of woodland caribou in relation to clear-cut logging in west-central Alberta. *Canadian Journal of Zoology* 78:1433–1440.

Sokal, R. R., and F. J. Rohlf. 1981. *Biometry.* 2nd ed. San Francisco: Freeman.

Solberg, E. J., A. Loison, B. E. Sæther, and O. Strand. 2000. Age-specific harvest mortality in a Norwegian moose *Alces alces* population. *Wildlife Biology* 6:41–52.

Stearns, S. C. 1992. *The evolution of life histories*. Oxford: Oxford University Press.

Steele, B. M., and J. T. Hogg. 2003. Measuring individual quality in conservation and behavior. 243–270 in *Animal behavior and wildlife conservation*, ed. M. Festa-Bianchet, M. Apollonio. Washington, DC: Island Press.

Steen, W. J. v. d. 1983. Methodological problems in evolutionary biology, I: Testability and tautologies. *Acta Biotheoretica* 32:207–215.

Stelfox, J. B., 1993. *Hoofed mammals of Alberta*. Edmonton: Lone Pine Publishing.

Stevens, V. 1983. *The dynamics of dispersal in an introduced mountain goat population*. Ph.D. thesis, University of Washington, Pullman.

Stevens, V., and D. B. Houston. 1989. Reliability of age determination of mountain goats. *Wildlife Society Bulletin* 17:72–74.

Stockwell, C. A., G. C. Bateman, and J. Berger. 1991. Conflicts in National Parks: a case study of helicopters and bighorn sheep time budgets at the Grand Canyon. *Biological Conservation* 56:317–328.

Swenson, J. E. 1985. Compensatory reproduction in an introduced mountain goat population in the Absaroka Mountains, Montana. *Journal of Wildlife Management* 49:837–843.

Swihart, R. K., H. P. Weeks, A. L. Easter-Pilcher, and A. J. DeNicola. 1998. Nutritional condition and fertility of white-tailed deer (*Odocoileus virginianus*) from areas with contrasting histories of hunting. *Canadian Journal of Zoology* 76:1932–1941.

Taillon, J., D. G. Sauvé, and S. D. Côté. 2006. The effects of decreasing winter diet quality on foraging behavior and life-history traits of white-tailed deer fawns. *Journal of Wildlife Management* 70:1445–1454.

Tavecchia, G., T. N. Coulson, B. J. T. Morgan, J. M. Pemberton, J. Pilkington, F. M. D. Gulland, and T. H. Clutton-Brock. 2005. Predictors of reproductive cost in female Soay sheep. *Journal of Animal Ecology* 74:201–213.

Terry, E. L., B. N. McLellan, and G. S. Watts. 2000. Winter habitat ecology of mountain caribou in relation to forest management. *Journal of Applied Ecology* 37:589–602.

Testa, J. W., and G. P. Adams. 1998. Body condition and adjustments to reproductive effort in female moose (*Alces alces*). *Journal of Mammalogy* 79:1345–1354.

Thomas, C. D., A. Cameron, R. E. Green, M. Bakkenes, L. J. Beaumont, Y. C. Collingham, B. F. N. Erasmus, M. F. de Siqueira, A. Grainger, L. Hannah, L. Hughes, B. Huntley, A. S. van Jaarsveld, G. F. Midgley, L. Miles, M. A. Ortega-Huerta, A. T. Peterson, O. L. Phillips, and S. E. Williams. 2004. Extinction risk from climate change. *Nature* 427:145–148.

Thomas, D. C., S. J. Barry, and H. P. Kiliaan. 1989. Fetal sex ratios in caribou: maternal age and condition effects. *Journal of Wildlife Management* 53:885–890.

Thomas, D. C., and D. R. Grey, 2002. Update COSEWIC Status Report on the woodland caribou *Rangifer tarandus caribou* in Canada. In COSEWIC assessment and update Status Report on the woodland caribou *Rangifer tarandus caribou* in Canada. Ottawa: Committee on the Status of Endangered Wildlife in Canada, 98.

Thompson, K. V. 1995. Flehmen and birth synchrony among female sable antelope, *Hippotragus niger*. *Animal Behaviour* 50:475–484.

Thompson, R. W., and J. C. Turner. 1982. Temporal geographic variation in the lambing season of bighorn sheep. *Canadian Journal of Zoology* 60:1781–1793.

Toïgo, C., J.-M. Gaillard, D. Gauthier, I. Girard, J. P. Martinot, and J. Michallet. 2002. Female reproductive success and costs in an alpine capital breeder under contrasting environments. *Ecoscience* 9:427–433.

Toïgo, C., J.-M. Gaillard, and J. Michallet. 1997. Adult survival of the sexually dimorphic Alpine ibex (*Capra ibex ibex*). *Canadian Journal of Zoology* 75:75–79.

Toïgo, C., and J. M. Gaillard. 2003. Causes of sex-biased adult survival in ungulates: sexual size dimorphism, mating tactic or environment harshness? *Oikos* 101:376–384.

Toïgo, C., J. M. Gaillard, and J. Michallet. 1999. Cohort affects growth of males but not females in alpine ibex (*Capra ibex ibex*). *Journal of Mammalogy* 80:1021–1027.

Trivers, R. L., and D. E. Willard. 1973. Natural selection of parental ability to vary the sex ratio of offspring. *Science* 179:90–92.

Van Ballenberghe, V., and W. B. Ballard. 1994. Limitation and regulation of moose populations: the role of predation. *Canadian Journal of Zoology* 72:2071–2077.

VanderWerf, E. 1992. Lack's clutch size hypothesis: an examination of the evidence using meta-analysis. *Ecology* 73:1699–1705.

Veitch, A., E. Simmons, M. Promislow, D. Tate, M. Swallow, and R. Popko. 2002. The status of mountain goats in Canada's Northwest Territories. *Proceedings of the Biennial Symposium of the Northern Wild Sheep and Goat Council* 13:49–62.

Verme, L. J. 1985. Progeny sex ratio relationships in deer: theoretical vs. observed. *Journal of Wildlife Management* 49:134–136.

von Hardenberg, A., B. Bassano, M. Z. Arranz, and G. Bogliani. 2004. Horn growth but not asymmetry heralds the onset of senescence in Alpine ibex males. *Journal of Zoology* 263:425–432.

von Hardenberg, A., B. Bassano, A. Peracino, and S. Lovari. 2000. Male alpine chamois occupy territories at hotspots before the mating season. *Ethology* 106:617–630.

Wade, M. J., and S. Kalisz. 1990. The causes of natural selection. *Evolution* 44:1947–1955.

Wauters, L. A., S. A. d. Crombrugghe, N. Nour, and E. Matthysen. 1995. Do female roe deer in good condition produce more sons than daughters? *Behavioral Ecology and Sociobiology* 37:189–193.

Weatherhead, P. J. 1986. How unusual are unusual events? *American Naturalist* 128:150–154.

Wehausen, J. D. 1995. Fecal measures of diet quality in wild and domestic ruminants. *Journal of Wildlife Management* 59:816–823.

———. 1996. Effects of mountain lion predation on bighorn sheep in the Sierra Nevada and Granite Mountains of California. *Wildlife Society Bulletin* 24:471–479.

White, G. C., and R. M. Bartmann. 1998. Effect of density reduction on overwinter survival of free-ranging mule deer fawns. *Journal of Wildlife Management* 62:214–225.

White, R. G. 1983. Foraging patterns and their multiplier effects on productivity of northern ungulates. *Oikos* 40:377–384.

Williams, J. S. 1999. Compensatory reproduction and dispersal in an introduced mountain goat population in central Montana. *Wildlife Society Bulletin* 27:1019–1024.

Wittmer, H. U., A. R. E. Sinclair, and B. N. McLellan. 2005. The role of predation in the decline and extirpation of woodland caribou. *Oecologia* 114:257–267.

Wright, G. J., R. O. Peterson, D. W. Smith, and T. O. Lemke. 2006. Selection of Northern Yellowstone elk by gray wolves and hunters. *Journal of Wildlife Management* 70:1070–1078.

Yoccoz, N. G., A. Mysterud, R. Langvatn, and N. C. Stenseth. 2002. Age- and density-dependent reproductive effort in male red deer. *Proceedings of the Royal Society of London B* 269:1523–1528.

Index

257